TROPICAL PLANTING AND GARDENING FOR SOUTH FLORIDA AND THE WEST INDIES

TROPICAL PLANTING

AND

GARDENING FOR SOUTH FLORIDA

AND

THE WEST INDIES

By

NIXON SMILEY

UNIVERSITY OF MIAMI PRESS

CORAL GABLES, FLORIDA

Fourth Printing, 1970

Library of Congress Catalog Card Number 61-9185

ISBN 0-87024-008-0

Since the publication of this book, many chemical preparations used in gardening and recommended in this book have been found to be unsafe for general use; among these are such previously accepted pesticides as chlordane, DDT, lindane, and toxophene. As research progresses, other chemical preparations may become regarded as dangerous. Do not use chemical preparations without first checking with an authority—for example, your county agricultural agent—who has on hand the latest information on such pesticides.

Manufactured in the United States of America

INTRODUCTION

Where to get information and how to use it are major problems for the south Florida gardener. The newcomer is baffled by the plants, the climate, and the soils. It is almost like moving to a new world. New facts must be learned. New adjustments must be made.

Providing the facts and helping to make the adjustments as easily as possible is the aim of this book. The facts in it are basic. They are put as simply as the author knows how. If some parts seem complicated, we ask the reader to bear with us. For we must confess that we have not found gardening a subject which can be so simplified that it can be learned without effort. It is like learning anything else. What you get out of it will depend on the effort you put into it.

A book on gardening should be readable and useful. This book was written with the hope it will serve both purposes. A quick reading should acquaint the gardener with its contents. Thereafter it should serve as a guide, providing answers to gardening problems which come up from time to time. But books alone don't make a good gardener. Experience is necessary.

One of the quickest ways to learn tropical plants is to visit tropical gardens, where many species are growing together. Florida and the West Indies are fortunate to have many fine botanical gardens. Those within easy reach should be visited again and again.

Dade County's own Fairchild Tropical Garden, at 10901 Old Cutler Road, has one of the largest collections of tropical plants of any other garden in the Western Hemisphere. Other excellent local collections may be seen at the Gifford Arboretum at the University of Miami; at the Redland Fruit and Spice Park, Coconut Palm Drive and Redland Road in south Dade County; at the Subtropical Experiment Station, on Waldin Drive, one-half mile west of Redland Road in south Dade County, and at the U. S. Plant Introduction Garden, on Red Road three miles south of South Miami.

Many fine botanical gardens also may be visited in the West Indies. Comparable with the best are Harvard University's Atkins Gardens at Cienfuegos, Cuba, and Hope Gardens at Kingston, Jamaica. Others are the Royal Botanic Gardens, Port of Spain, Trinidad, and the Botanic Gardens of the Imperial College of Agriculture, Dominica. Smaller gardens may be seen at the islands of Grenada, St. Lucia, and St. Vincent. Although no public botanical gardens exist in the Bahamas, some outstanding private gardens have been developed at Nassau, into which the visitor who is genuinely interested in tropical plants usually has no trouble gaining entry.

Thus through reading books and bulletins, by visiting gardens, and at the same time engaging in the practice of gardening, one should learn quickly. It is hoped that *Tropical Planting and Gardening for South Florida and the West Indies* will provide the information needed by the gardener whose main desire is to create pleasing surroundings in which to live.

CONTENTS

Chapter 1

CLIMATE AND THE GARDENER

It is important for the gardener to know something about the climate where he lives and to appreciate the effects of climate on the plants he grows.

Gardening in the tropics or the subtropics is, in many ways, similar to gardening in cooler latitudes. Plants require moisture, nitrogen, minerals, and light; and they must be kept relatively free of insect pests and diseases. But while gardening in the cooler areas is done only during part of the year, in the warmer areas gardening tools are never stored away for a season. For grass must be mowed, pruning may be needed throughout the year, and insects never rest.

Climate is variable everywhere, whether in the tropics or in the colder zones. Altitude or prevailing breezes may affect the climate. The Torrid Zone, noted for its hot and humid conditions, also has the world's most agreeable climate. It depends on where you are. In Quito, Ecuador, only sixteen miles from the equator, the climate is never excessively hot or cold. The city sits in a high valley in the Andes, 9,500 feet above sea level. The climate of Guayaquil, Ecuador, less than an hour's flight from Quito, is greatly different. Guayaquil is situated on a river in a flat country near the Pacific, only a few feet above sea level. Here the temperature and the humidity are high, a combination which makes the climate disagreeable. The climate of many areas in the Temperate Zone, frigid in winter, may be just as hot and humid in summer as in Guayaquil, which is in the Torrid Zone.

One finds a comfortable climate in the highlands of Mexico, Guatemala, Honduras, and the Andes. The mild climate of northern Europe is credited to the influence of the Gulf Stream. The pleasant climate of Honolulu and that of some of the West Indian islands is said to result from favorable breezes.

The plants in your garden are affected by climate, just as you are. But whereas you can seek shelter from a hot sun or enjoy the warmth of a heated building during cold weather, the plants in your garden are at the mercy of the heat, the cold, the wind, and the rain. For your plants to thrive they must be adapted to the climate prevailing in your garden throughout the year. A plant adapted to the north, where ice and snow cover the ground much of the winter, is not likely to thrive in Florida gardens, nor should one expect to see an Amazon valley palm growing well in Boston Common.

Apples, peaches, and pears are examples of plants which require periods of winter chilling in order for them to bloom and to set fruit. But there is great variability among varieties of these. A peach variety which thrives in Ohio may not fruit well in Georgia, for it probably would require a longer chilling period than it would receive in the South. Several varieties of peaches can be grown in north Florida which would be worthless to plant in south Florida. These same peach varieties probably would be killed by freezing in Illinois.

The climate of south Florida is too warm for growing the deciduous fruits, as apples and peaches are classed, and the planting of them here is not recommended. But there is always that inevitable exception to the rule. The Ceylon peach, introduced to Florida from the Far East seventy-five years ago, fruits well in south Florida. It apparently does not require a chilling period to induce it to bloom and fruit, or, if a chilling period is required, it must be negligible.

Other variables in climatic effects also must be considered. For instance, few of the desert plants of Africa or the Southwest would thrive in Miami gardens. Nor would most of the high altitude plants of Guatemala or the Andes. On the other hand, there are many plants which are so tropical in their demands that they cannot survive Miami's mild winters, even when the temperature may reach no lower than 45 degrees. Plants from tropical rain forest areas may do well in south Florida only under artificially moist conditions. For part of south Florida's year is dry and cool and part of the year is hot and wet. A plant that requires year round high humidity would suffer during our dry season. The climatic requirements for the world's many species of plants is varied to no end. The gardener will do well to select those plants which have proved their adaptability here.

Temperature, Not Latitude, Most Important

Where is the line that divides the tropics from the Temperate Zone? And what is the difference between the tropics and the subtropics? Early navigators and map-makers considered the tropics as being in the Torrid Zone, a belt lying between the Tropic of Cancer and the Tropic of Capricorn, and which was bisected by the equator. The Tropic of Cancer is the northernmost line reached by the overhead sun in summer and the Tropic of Capricorn is the southernmost line reached by the overhead sun in winter. The Tropic of Cancer crosses the Florida Straits between Key West and Cuba and passes through the Bahamas south of Nassau. Havana is located just inside the Torrid Zone, while both Miami and Nassau are in the Temperate Zone.

Actually there is no sharp line dividing the tropics from the Temperate Zone, so far as weather is concerned. Arctic and Antarctic frigid waves sometimes invade the Torrid Zone. Tender tropical plants may be severely injured or killed. When invading cold waves are severe, even large trees may be killed. On the other hand, when the sun moves to its northernmost overhead point in summer, it is closer to New York than it is to the equator. Is it any wonder, then, that Temperate Zone states get so hot in summer?

It is the kind of climate a place has throughout the year that determines what kinds of plants may be grown. Miami is usually thought of as being in the subtropics. It is influenced more by the climate of the tropics than by the climate of the north. Its ornamental plants and its fruit trees are mainly of tropical origin. Northern types of vegetables and annuals are grown in Miami, but only during the cool months.

Nassau, on the other hand, is usually thought of as being in the tropics. Although most of the Bahamas lie above the Tropic of Cancer, these islands never receive any frost, and a number of very tender tropical plants can be grown in Nassau which cannot be grown outdoors in Miami. Nassau has large breadfruit trees which are never harmed by cold. In Florida, only in Key West does one see mature breadfruit trees. On the mainland breadfruit trees are subject to periodic killing-back by frost. In the outskirts of Miami, near the Everglades, a breadfruit tree would be killed to the ground nearly every winter.

The lowest temperature recorded in Miami was 27 degrees, during the 1917 freeze, one of the most severe freezes in the history of south Florida. The temperature dropped below freezing soon after sundown and remained below freezing until mid-morning of the following day. Such cold was too severe for a great many tropical plants, which were killed to the ground. Fortunately south Florida seldom gets such hard freezes. There was not a severe freeze between 1940 and 1958. The 1958 freeze was bad but not nearly so bad as the 1917 freeze. Recovery of stricken plants was rapid and within a few months hardly any signs of cold damage remained. But if south Florida had such freezes every winter its tropical plants would not be able to make full recovery from one winter to the next. And soon the cultivation of the more tender tropical things would cease.

Horticultural Limits of Tropics

Plant scientists have offered a number of suggestions about how to determine whether an area is tropical. The northernmost limit of the fruiting coconut palm has been considered by some to be the northernmost line of the tropics. This would put most of south Florida in the tropics. Others have suggested that only any area where a breadfruit tree can be grown to maturity should be considered tropical. The idea of using the breadfruit tree as a guide for determining the limits of the tropics seems practical. Where a breadfruit tree can be grown to maturity, almost any other tropical plant also can be grown, if moisture and soil conditions are favorable.

The northern limit for the cultivation of oranges is often considered to be the northern line of the subtropics, It is often said that Miami is on the edge of the tropics. Most tropical plants can be grown here, but few northern species thrive here.

Plants are just as variable as the weather. It is often difficult to explain to newcomers to south Florida why they cannot grow apples, hydrangeas, tulips, and blue grass, such as they grew back home. Many newcomers, missing the plants they have been associated with all their lives, insist on bringing these plants with them to Florida. It is hard for them to believe that some plants do require cold winters where snow and ice is common and where the ground may freeze some years to a depth of two or three feet. They reason that if the climate of south Florida is so favorable for plant growth, as it is claimed, then the plants which thrive in New York or in Nova Scotia should do even better where the weather is favorable throughout the year.

Plants become adapted to a climate through centuries of exposure. The winters of New Jersey or Pennsylvania may be too mild for many species of trees native to the country north of the White Mountains. Likewise, the climate of south Florida is too warm for most of native plants of north Florida. The principal reason why West Indian plants dominate the hammocks and keys of south

Florida is due mainly to the fact that the climate is more favorable to their growth than to the growth of Florida's Temperate Zone plants. As one progresses northward, the Caribbean flora gives way to the Temperate Zone flora.

Plants do vary in their adaptability. The live oak seems to be as much at home in south Florida hammocks, where it is almost the lone representative of the Temperate Zone flora, as it is in north Florida or Virginia. The wax myrtle grows from the southern tip of Florida to New England. The slash pine is at home from South Carolina to the Caribbean Islands. There is good reason to wonder, though, whether a slash pine seedling from the Isle of Pines, Cuba, would grow well in north Florida, or whether the north Florida slash pine would thrive at the Isle of Pines. Foresters have found some difference in the adaptability between the south Florida slash pine and the north Florida slash pine. Some differences of adaptability to climate may exist in varieties of live oak and wax myrtle.

A few plants may be moved, within limits, from one climate to another. *Ligustrum japonicum* and *Pittosporum tobira* seem to be almost equally at home in Miami or in Jacksonville, 350 miles to the north. Some varieties of fire thorn, or pyracantha, appear to thrive as well in Miami as they do in Raleigh, North Carolina.

There is no easy way to explain the differences in plant adaptability to climate and to soil. The gardener endowed with an experimental nature will do well to use plants for basic development of a garden which have been tested and proved in his area. The use of untried ornamentals, as well as delicate plants which require protection from cold or beating winds, should wait until the basic planting in the garden has been developed.

A MacArthur palm, espaliered Clusia, clinging Ficus on old brick wall, bayonet-leaved Nolina, and Liriope ground cover—all superbly planted against a background of live oaks—make this tropical garden spot.

Chapter 2

DESIGNING THE TROPICAL GARDEN

The way a garden is landscaped and planted may well determine whether one lives in a distinguished home or in an ordinary house. As much thought should be given to garden design and to the selection of trees and shrubbery and their planting as is given to the design and the building of the house. The garden should be planned as the house is planned. First, the existing native growth on the lot should be considered. The shape and placement of the house may be determined by the position of existing trees. Where many trees and native shrubs exist, the lot should be carefully hand-cleared, rather than cleared by a bulldozer, in order to avoid damage to the plants. If landscape plans call for the planting of one or more fairly large shade trees in the rear of the lot, it may be necessary to do the planting before the house is begun. It may be impossible to get a tree truck into the rear of a small lot after the house is built.

It pays to have the services of a landscape architect, not only to design the garden but also to give advice about preserving native plants. Many of the wild, scrubby plants which are bulldozed off building lots may be superior to the exotics which are planted in their places. Native plants are adapted to the soil in which they are growing. It may be a good idea to save many of these plants, including some of the pineland saw palmetto. But what to save, how much to save—or whether to save anything—will always depend upon circumstances. This is where expert advice is helpful.

For instance, in some of the high pineland in Dade County the surface is so rocky that it is next to impossible to plant anything. Moreover, the existing plants may be no more than a few slender Caribbean pines, palmettos, and poor pineland scrub. It may be necessary to have the entire lot cleared by a large bulldozer and plowed to a depth of six or eight inches. With the aid of fertilizer and mulching it is possible to grow a large number of plants quite well in six inches of loose rock and sand lying over solid limestone. Nearly all of the avocado groves in south Dade County are grown under such conditions.

If one likes pines, and the trees on one's lot are especially fine, then the use of a bulldozer should be restricted to the actual building site and whatever open area one intends to have in the front and in the rear. Unless pines are growing in deep sand they may have their roots damaged severely by the bulldozer blade, and, being so weakened, are likely to die within a few years. Uneven surfaces about pines can be filled in and leveled with sand. Neither marl nor muck should be used.

Landscape Plans Come First

There are several other reasons for giving so much attention to the planting of the grounds before a house is built. What kind of trees and shrubs does the owner want? Is he really interested in gardening? Is he interested in a certain group or groups of plants, such as palms, hibiscus, mangos, avocados, lush shade plants such as philodendrons or anthuriums, or in building an orchid collection? Planning, planting, and maintaining even the most simple kind of garden requires some knowledge of plants. A garden containing a great many features, and especially a garden in which there is a lot of color, may demand almost constant attention. The owner must have enough interest and time to give some effort nearly every day to the garden. Since this is seldom the case, because most persons have a wide variety of interests besides the growing of plants and maintaining a garden, it is always best to start out with a very simple garden—well planned, but containing a minimum of planting of materials well adapted to the soil and to the climate.

Here again, if the planning of the garden is done as plans are drawn for the house itself, it may be possible to begin collecting planting materials months ahead of the time when they will be needed. Where one has small funds to invest in planting, it is often possible to save money by buying plants before they will be needed, and to transfer them to larger containers and so push their growth into larger specimens.

Only tough, adaptable types of shrubs and trees should be selected for the basic planting. What specific kinds chosen will depend to a great extent upon specific needs; but too often selections depend upon the desire of the owner for a quick, colorful garden. Gay-toned plants, colorful ground covers, beds of annuals and the like may have their places in the garden, but the basic planting should be done first.

So many persons look upon the tropics as a land of perpetual color. This is hardly a true picture. The principal color in the tropics is green. One sees bright colors here and there, but green is everywhere. The garden should be planned with this in mind, with green dominating and bright color playing a secondary part.

Every person who plans a garden must have some ideal picture in mind; but this is usually vague and difficult to define. The problem to make it real. The dream of the ideal is difficult enough to put down on paper, even by the best landscape architect, and to plant and maintain the ideal requires the knowledge and the skill of an outstanding gardener.

Planning Must Be Realistic

Most of us are more likely to come nearer to perfection if we put our conception of the ideal garden in the background and tackle the problem from a practical and realistic standpoint. There is the neighborhood to be considered. How much screening must be provided and what kind of plants must we use? Some consideration must be given to those cold northwest winds that often strike south Florida in the winter. It is possible to provide some protection against the winds, and what kinds of plants might we use? Can we arrange plantings so that we might enjoy the southeast breeze in summer? Are there many children in the neighborhood and should we consider erecting a steel fence before putting in any plants?

All such decisions should be made in advance if possible. But no matter what, the main plantings should be restricted to the borders and the first line of plants on the outside should be of the most adapted species—plants which, when well established, will be capable of withstanding cold, wind, drought, and the full effects of the summer sun. The kinds of plants suitable for such basic planting are discussed under *Shrubs and Their Uses.* Haphazard planting, the sticking of plants here and there throughout a yard, is the worst enemy of good design. It steals space and prevents full use and enjoyment of a building lot, which, in most instances, is small enough anyway. By planting here and there we make the area much smaller than it really is. Such planting is completely lacking in artistic merit, and gives one the impression of a mess rather than of a garden.

In a well planned garden one always finds a great deal of open space. On large estates it is possible to go from one open area to another, connected by trails through naturally wooded areas or through masses of especially planted trees and shrubs. But in the back yard of a small city lot it is hardly possible to have more than one open area, which becomes more or less the center of the garden. There is usually room for only one or two shade trees, the placement of which will depend to a great extent on what special features are planned. Trees should not be planted where they will drop their leaves into a swimming pool, nor should they be planted so close to the house that large branches will eventually spread over the roof to create excessively moist conditions during the rainy season. Nor should any tree be planted so close to the lot line that its branches spread far over on to the lot of a neighbor. Trees that become huge, such as rubber trees, create dangerous situations when they reach maturity. One *Ficus benjamina* is capable of covering most of the ground of the back yard of an average city lot. Its roots are aggressive, spreading out under neighboring lots and sometimes traveling as far as two hundred feet to invade neighbors' septic tank drains. These huge trees create serious hazards during hurricanes, because their branches, of tremendous size and weight, are quite brittle and are usually broken up by high wind. A number of more appropriate trees for the small garden may be found in the section *Trees and Their Uses.*

Make Front Yard Simple

Planting of the front yard should be just as simple, or even more so, than the back yard: We must forget the landscape scenes we like so well in parks and about formally planted estates. A home is not a tourist attraction; it is a place to live in relaxed surroundings. It is not a reception hall for a monarch. Formality in planting should be avoided, but we do not mean to suggest that planting be done haphazardly. A few well placed shrubs and trees in the front yard will do more for the appearance of a house than scores of hibiscus, ixoras, acalyphas, and crotons, which clutter so many front yards. It is all right to use some colorful plants, but they should play a secondary part in the planting. Just as with the back yard garden, only really tough plants should be used for hedges, screens, and in other basic plantings, use only long tested shrubs and trees.

Actually there is little opportunity to make an outstanding garden in the modern front yard. A city lot is small anyway, and after space for driveways is taken out very little room is left for grass, trees, or shrubs. The automobile is a part of any home grounds today. Thus the planning of the driveway deserves a great deal of attention. Wherever possible, there should be ample space in a front yard to turn an automobile around. Straight driveways that lead from a street to a carport or garage, forcing one to back a car out into the street, create a hazard and should be avoided if possible. The importance of correct planning of driveways cannot be over-emphasized.

Each home has its own problems. The landscape architect, who also is trained as an engineer, should be given the job of laying out the driveway so that it may go well with the house and grounds as well as being functional. The asphalt concrete now being used to make driveways soon bleaches under the sun and loses the black color which makes it look so objectionable at first. For concrete driveways, green color can be mixed with the finish, so that it blends in with the surroundings. Paved driveways, when expertly planned, can save the homeowner a great deal of time from lawn care. In fact, it is possible to plan the driveway in such a way that grass need not be planted in the front yard.

If the driveway is to be circular, the center can be filled with a ground cover, such as wedelia or jasminum. Or, it may be given over to a few slender palms or to several small trees, with the ground covered with brown river gravel. Shrubbery may be massed along the borders of the lot, and low ground cover types of plants may be used in the remaining space. Or gravel can be used close to the driveway to give persons getting in and out of automobiles a place to walk.

There are many solutions to the front yard problem. But first we must get away from conventional ideas. Why do we need grass in front of the house? Why do we need beds of flowers? Who are we trying to impress? Why not plan first for comfort and safety? We could surely find a little space to put in a few plants suitable for the location. There will always be room for a hedge or a few screening plants. There should be ample room close to the house for a few well chosen foundation shrubs. There may be room for one or two small trees in the front yard. Above all, we should avoid the planting of twins in the front yard, such as two hibiscus on either side of the walk, two sheared bougainvilleas, two conical arbor vitae, two palms, two ixoras, or two crotons. We should avoid lining driveways or walks in front of small homes with palms or other plants. What purpose does such planting serve? Why line driveways and walks with low formal hedges? Why plant huge ficus trees in the front yard to overwhelm the house? Why fill the front yard with numerous kinds of shrubs which must be kept rounded by the pruning shears of commercial gardeners? Simple plantings and an adequate, open driveway, are especially suited for modern living. The automobile most certainly deserves a place in any plan for the front yard, and that not only goes for our own automobile but the automobiles of our friends who come to visit us.

Outdoor Living

So far we have discussed only garden planning in general. No thought has been given to special features. Yet, nearly every garden has one or more of them: a swimming pool, a patio, outdoor cooking area, play area for children, a rock garden, lily pond, vegetable or annual garden, or a tropical fruit collection. Nor should we forget the hobbyist, such as the rare plant collector, or the orchid grower. These special features are often developed after the main garden is designed and planted. It is helpful in selecting planting materials for screening and background if a rough idea of future plans can be known in advance. Even if a swimming pool or patio is to be added far in the future, some indication of the location should be made while the house and garden plans are made.

How to build lily ponds, construct walks or how to lay out swimming pools is beyond the scope of this book. Books on garden architecture may be found in public libraries or inexpensive ones bought which illustrate examples of garden construction, including slat houses, tool houses, outdoor cooking areas, and patios. We should caution here that many of the structures illustrated in these books, and in many magazines, are not acceptable under Florida zoning laws. The zoning laws of some cities and counties forbid the construction of slat houses or glass houses. The light plywood storage houses and work sheds, often seen illustrated in magazines, become dangerous hazards in hurricanes, because they quickly blow to pieces and their parts are scattered over the neighborhood. Plywood is strong, but it must be used according to building code requirements so that structures made of it do not blow apart during storms. Outdoor furniture, fences, plant boxes and other things exposed to the weather do not last long in Florida unless made of pressure-treated wood or concrete. Redwood may last for several years if it does not come in contact with the ground. But even treated wood eventually decays.

In general, it is best to avoid unusual designs. Circular stepping stones running this way and that way through the grass from the street to the front door may be rather unusual, but is this kind of walk practical? Perhaps in time we might get tired of these circles, especially after we have to edge the grass from around them for a summer. If circular stepping stones must be used, it is advisable to consult a landscape architect and perhaps get him to place them.

The same suggestion might be made for winding walks. Few things are more ridiculous than a walk which winds through the grass of one's front yard. Why the curve? What is the curving walk avoiding? Certainly not the grass. Perhaps there may be an occasion when a curving walk through a green lawn may be appropriate, but ordinarily only an artist could lay out such a design. The same is true of free-form swimming pools. Such pools can be quite practical when built in a patio and laid out by an expert. But it must appear to have a reason for its existence. If there is no reason, then a rectangular pool might look much better. In general, a round or rectangular lily pool is much to be preferred to the so-called informal pool, laid out ostensibly to resemble a portion of a winding stream. The winding stream idea may be all right, if we can cope with it. Usually the pool looks contrived. If it fails to give the impression of a part of a winding stream, it may look ridiculous.

A conventional garden pool is usually quite satisfactory, especially if we make it large enough and deep enough to plant with colorful water lilies. Filling a pool with all kinds of aquatic plants should be avoided, unless one really has a fairly large bog pond which looks like a bog pond and which contains enough soil, or mud, for the growth of bog plants.

Plants Should Seem Right

When we try to imitate nature, such as making a desert garden, or a bog pond, considerable restraint must be exercised. What could be more out of place than a hibiscus or a croton in a desert garden, or an agave or yucca growing on the edge of a bog pond? Here we are dealing with a subject known as ecology. Plants in nature form their own communities, and these communities vary throughout the world according to the climate and the soil. We don't find bog plants growing in deserts, nor do we find desert plants growing in swamps. If we go in for special gardens which we might call "natural gardens," then we must do our best to make them appear natural. There is nothing wrong with the use of desert type plants from Africa, Arizona, and Australia, all in the same sandy area. Likewise there is nothing wrong with putting bog plants from all over the world in a bog pond.

Appropriate background is just as important as community association, because it is hard to separate plant communities from the background. The dry garden, for instance, should be removed from any lush tropical background. The yucca makes a satisfactory background for it. Sometimes dry gardens are built against a wall, such as a south or west wall within a patio where both direct and reflecting sunlight make it very difficult to grow other things. Broadleaf evergreens might form the background for the bog pond. But sometimes a planting of native scrub willow, as seen growing around Everglades ponds, may be appropriate if the area is large enough. These willows lose their leaves for a few weeks during the winter, but when their new foliage comes out in February or March it remains a bright new yellow-green for several weeks, making a striking contrast against the darker background foliage. Another effective plant for a pond border is the native cocoplum. It should be planted where its sprawling branches will dip into the water. But it must be remembered that the cocoplum eventually grows into a large shrub, and is not suitable for a small pond. But here we get into scale, one of the most important elements of design, and which we will discuss.

Study of Design Important

The theory of design covers all forms of art, including flower arranging, painting, architecture, landscape architecture, and the laying out of super-highways. Many books are available on design theory, covering the subject from the most simple approach to the most abstract conceptions of the various schools of design. Unfortunately, most home owners have no educational background in design, which, like botany, is not taught in public schools. So many persons know little about either design or about plants that they have no basis for realizing how little they do know. When we are completely lacking in knowledge of a subject, the more likely we are to look upon it as a simple thing, easy to master with a little effort. But garden designing, like any other kind of designing, is not easy even for experts. Nor is gardening easy. Both designing and gardening take years of study and experience if one is to master them. Thus the home owner, who generally tries to cope with the problems of designing and planting a garden faces a doubly difficult situation. About all we can hope to accomplish here is to make the reader appreciate the need for further study. Here is an outline of the features of garden design:

COMPOSITION is the arrangement of all features in a garden, including plants and structures, so that a pleasing relationship is attained. Here we have an interplay of space, form, texture, and color. How to achieve good composition requires knowledge of scale, balance, accent, rhythm, and contrast.

SCALE is the relationship in size of parts of a plan. A huge rubber tree dominating a small house is obviously out of scale. Large trees go with large houses. The same is generally true of palms. Medium sized trees and palms should be selected for small homes on city lots. Thus we immediately find ourselves getting into a problem where both knowledge of design and knowledge of plants come in handy.

BALANCE is also a relationship between parts of a plan, but it is a quality felt rather than to be measured. Something tells the mind whether the elements in a landscape scene are in balance. Balance may be formal or informal. Many home owners try to achieve formality by placing identical palms or arbor vitae on either side of an entrance, or identical coconuts on either side of a walk near the street. Elementary formal balance may be easy to obtain in an amateurish way, but fine formal gardens are designed only by experts.

Informal balance is more difficult to achieve. Woodland scenes which give us so much enjoyment are examples of the way informal balance is obtained in nature. Masses of trees in one part of a scene balance with hills, glades or islands of trees in another part of the scene, forming, altogether, a balanced and pleasing picture. When the landscape architect masses several large shrubs on one side of a garden to offset the effects of a single spreading tree on the other side, he is striving for the same kind of balance one sees in nature. The inexperienced gardener may sense something is wrong in a garden in which there is a lack of balance, but his lack of knowledge of design may prevent him from solving his problem.

ACCENT is used to break monotony in the balanced landscape. It can be a plant, a gate, a rustic seat, a pool, or a fountain.

RHYTHM is the repetition or a sequence of lines closely tied to the element of balance. Rhythm can be obtained with steps, paving blocks, or benches, as well as in the proper placing of plants of similar color and texture in relation one to another. How to create rhythm may be difficult for the amateur gardener. But we might offer one illustration which might help: if a patio is made of large rectangular stepping stones, adjoining plant beds are likely to seem right if made of the same rectangular form.

CONTRAST can be achieved through the use of texture, color,

form, and shadow, as well as in the comparative heights of plants. In most instances contrast is an accidental result of garden planting and development rather than planning.

COLOR plays an important role in all gardens. It should be used with restraint. When richness is overdone, all the other elements of design are lost, and one winds up with glaring monotony.

TEXTURE is a more subtle element, but plays an important role in the gardens of sensitive designers. Foliage texture can relieve the monotony of plant masses. Transition between foliage textures creates contrast, and even may add a sense of rhythm through repetition. Thus shrubs should be selected as much for the texture of their foliage as for their form or the color of their flowers.

LIGHT and SHADOW, like color and texture, are elements that can be used to add interest to a garden. And they are the only elements that are constantly changing, as the sun moves. Examples are the changing shadow of a yucca on a wall, the lengthening shadows of shrubbery late in the day, or the sunlight playing on bright green foliage rustled by a breeze.

FORM and SPACE work together to give us the effect of three dimensions. Habits of trees and shrubbery must be known before they can be effectively grouped together, or planted in relation to one another. There are many plant forms, and every plant naturally takes on a certain form under varying conditions of growth. We must know which plants and which forms are best for the effect we wish to create. Forms are upright or sprawling, weeping or viney, round or explosive. It is through the use of the different possible forms in relation one to another that we create the strongest feeling of three dimensions. And it is in the use of these different forms that we combine all the elements of composition to attain pleasing garden design.

Drawing Plans to Scale

With the hope of doing a fair job, but with the realization that perfection may be elusive, the home owner should be ready now to go ahead with plans for the landscaping of his place. Details should be drawn to scale. This is not so difficult if graph or engineer's paper is used. Each small square can represent a square foot of ground. If one sheet doesn't have enough squares, paste several sheets together, or enough so that the full size of the lot can be laid out. Now draw in the plan of the house, walks, driveway, and also the location of the septic tank and drainfield, power and telephone lines. Next put in existing trees and shrubbery, as well as any large trees which hang over from adjacent lots. Draw an arrow to indicate the prevailing southeast breeze and another arrow to indicate the northwest wind of winter. Mark east and west so that sun and shadows are kept in mind as the sun moves overhead.

Now with the existing things indicated, it is time to make decisions about the design of the garden and the placement of trees and shrubbery. The foot-square areas of the graph paper should remind one that plants do cover space; that a tree which some day may be thirty feet across must be planted at least fifteen feet from the lot line if it is not to invade the air space over a neighbor's lot. And not even a small shrub should be set directly on the lot line.

The rest is up to the designer. He must decide on the kinds of plants, how many plants will be used, and how to place them. Every person, even the best designer, will tackle the problem differently. So the amateur, trying to lay out his first garden, should not worry about whether his plan will come out exactly like one by a landscape architect. One thing is sure. The amateur's time costs nothing, and graph paper is inexpensive. He can keep on designing until he comes up with something that is pleasing—at least to himself.

Many plants, like this ginger, are increased by division

A well maintained courtyard adds immensely to worth of home

Chapter 3

TROPICAL HORTICULTURE

Florida's soils are very poor. More fertilizer must be used here than in any other state. It is hardly possible to produce ornamentals or commercial crops without help from fertilizer. The soils of north Florida are poor enough, but the farther one travels southward down the peninsula the poorer the soil becomes. About all we can brag about in south Florida is the climate. What we have traditionally called soil is usually nothing but sand or limestone rock. Fortunately, this sand and rock can be made productive within a short time through the use of organic materials, mulching with leaves, grass clippings, wood shavings, peat moss, and any other plant particles which are subject to decay. Decay can be hastened through the use of fertilizer.

To understand how soil can be improved let us take a look at a tropical jungle. Matheson Hammock is a good example. We find the floor of the hammock covered with leaves. The scene is not greatly different from any other tropical jungle. Even most of the plants are of tropical origin, except the live oak which is represented here by several noble specimens. If we scratch down into the brown covering of leaves on the floor of the hammock, we get a picture of what has been happening for a very long time. Only the leaves on the surface still have their original forms. Deeper we find progressive decay until the dark particles can no longer be identified as leaves. Beneath this we find a dark sand mixed with a black material which, for all appearances, might be a muck. Digging a little further we come to what appears to be solid limestone. What we have are a few inches of soil and decaying organic matter lying on the surface of the rock, the result of many years of leaf fall and leaf decay.

Looking at this thin layer of humus a little closer, we find that it is filled with a network of roots of the many species of trees growing about, pigeon plum, mastic, red stopper, Spanish stopper, strangler fig, and ironwood. If a poisonwood tree is growing nearby, its roots also may be part of this network, and the hands of one digging into the humus may later break out with a rash, similar to the rash which develops after contact with poison ivy. The competition between roots is tremendous. They penetrate every nook and crevice in search of nitrogen and minerals. If a root dies another root immediately takes its place.

Soil is Full of Life

More takes place within this shallow humus than meets the eye. We may uncover an earthworm or an unfamiliar bug in our digging, but it is impossible for us to see the multitude of microscopic animals, bacteria, and fungi which live in this humus. Through their actions the organic matter is broken down and returned to the original soil solution from which it came. Thus new leaves are produced and in time they fall and join the debris on top of the soil already undergoing the processes of decay. Here the new leaves are attacked by microorganisms and in time are converted to humus. In the processes, nitrogen and minerals are released, as well as acids and other compounds, and these percolate through the overlying materials into the black soil, where they are absorbed by tree roots. This is a simplified picture of what takes place. The whole story is a complicated one which is not yet completely understood by scientists who have spent their lives studying such problems. But at least we have gained enough insight into these processes for us to be able to duplicate nature's methods in our own gardens.

The conditions which exist in Matheson Hammock may have been hundreds of years in the making. The owner of a new home on a raw subdivision lot cannot wait for the processes of nature to run their course. He needs to transform his poor sand or limestone into productive soil in the shortest time possible. For he has visions of his home being set off at once by a strikingly picturesque garden, with green lawn, and with borders of massive shrubbery, as well as shade trees, fruit trees, and perhaps beds of flowering plants. So often an attempt is made to develop this admittedly idealistic garden before any effort is made to solve the soil problems. With a few scrawny shrubs purchased at the lowest prices possible, and with as many cuttings as can be obtained from friends and neighbors, the garden is launched—with ninety-nine per cent hope and the remaining one per cent divided between effort and knowledge.

We discussed in the preceding chapter how to go about planning a garden. The next step is the planting; but one cannot expect shrubs and trees set out in poor soil to make much headway, even if they survive. We can overcome the soil problem to some extent by purchasing full-sized shrubs and trees, and keep these in good growing condition with the aid of fertilizers and ample moisture, with perhaps an occasional application of a nutritional spray containing zinc and manganese to correct mineral deficiencies. Most homeowners don't have money to plant mature shrubs and trees. They must start out in a small way with small plants. Thus the building of a good foundation for these small plants to encourage healthy growth is very important.

How to Make Compost

One of the first things the owner of a new house should do is to start a compost pile. One can find as many directions for making a compost heap as there are people who make them. The simplest way is to pile up grass clippings, leaves and any other organic particles obtainable. Scatter fertilizer over the pile periodically as it is being built and keep it moistened to promote bacterial action. A better way is to put down a layer of fresh organic materials about a foot deep. Add to this an inch of pit sand or builder's sand, plus half a dozen handfuls of fertilizer and several coffee cans of pulverized sheep manure or cow manure. Then add another layer of organic materials and sand and fertilizer, until you have a flattened heap five or six feet high. Cover as many square feet of ground area as may be needed for the amount of composting materials available. Compost heaps should not be permitted to dry out. Decay is brought about by microorganisms, which must have moisture for their complicated chemistry. Where there is no moisture, decay does not take place at all, and where there is little moisture decay is slow. Decay is most rapid in well moistened, well aerated compost heaps.

After several weeks, when decay near the bottom of the heap is well underway, the heap should be turned upside down. As soon as the materials decay enough to fall apart when picked up with a pitch fork they are ready for use. By now it will be noted that the heap is probably one third its original size. Compost is basic material for soil-building. It can be incorporated with the soil when shrubs or trees are planted, while the rougher particles of the material can be used for mulching after the planting is done.

Compost, in itself, does not contain enough nutrients, or does not release nutrients fast enough, for rapid growth of trees and shrubs. Additional fertilizer may be necessary to push young plants along toward maturity. But in the meantime you are building soil and creating ideal conditions for healthy plant growth. Just how rapidly such favorable soil conditions are developed will depend to a large extent on the gardener's own efforts. Unless humus is being continually replenished, it disappears after a few years, and the original poor conditions return. In a warm climate humus is broken down rapidly by microorganisms. This is why muck or peat disappears rapidly after it is applied to a lawn as a top dressing. We often hear it said that the "muck blew away." What happens is that the muck is broken down by microorganisms and the molecules returned to their original form, either entering the soil in solution or entering the atmosphere as a gas. Organic materials are made up mainly of carbon, nitrogen, oxygen, and minerals. When organic structures, or molecules, are broken down each of these elements goes its own way, and the original material which we recognize as peat, muck, or leaf particles disappears completely.

The processes through which this action take place are much more rapid in warmer areas of the world than in the cooler areas. The popular belief that the tropics are so rich in humus is largely myth. Organic material on the floor of a tropical jungle is broken down rapidly and disappears. The destruction of the humus is much more rapid after the jungle covering has been removed, as when land is cleared for agriculture. The humus completely disappears within a year or two, and the land is abandoned to return to jungle growth. After several years the soil is renewed, and once again is begun the cycle of clearing, cultivating, and abandoning.

In the cooler areas of the world the soil retains its humus much longer under cultivation than it does in the tropics. Contrary to common belief, humus may accumulate to a greater depth in northern forests, in grassy plains, and in bogs than in similar areas in the tropics. Most of our garden peat is obtained from the cooler areas of the world, where immense deposits have been created through partial decay of organic matter, especially certain mosses in swampy areas. These peat areas may eventually subside after being drained and cultivated; but in the warmer areas of the world where peat exists, it disappears rapidly following drainage and cultivation. A good example is what has happened in the Everglades, where rich peat deposits have subsided as much as eight feet in less than forty years of cultivation. Eventually the subsidence will be complete, if the present methods of cultivation are followed, and the area will have to be abandoned as an important farming center.

Thus, it is obvious that humus is never a permanent thing if it is used; and it certainly is of no value to the gardener unless his plants are making use of it. In older gardens, however, there need not be so much concern for the loss of humus, as it is being constantly replaced through the decay of roots of existing plants, and especially grass.

Humus is especially needed where gardens are made in limestone soil. The humus releases a constant supply of acid when the soil is moist, thus reducing the alkalinity and making available minerals that limestone soil tends to fix in chemical compounds which cannot be absorbed by plant roots. The acid may change these compounds into soluble forms which become available to plants. Thus the constant chemical changes that acids and bacteria bring about in humus-laden limestone soil offer to plant roots a never ending source of minerals.

On the other hand, in some areas of Florida the soil is highly acid, especially in Florida's palmetto and gallberry flat woods. While many plants thrive in acid soil, especially plants like ixora, whose foliage turns yellow when planted in limestone soil, other plants may not thrive under such highly acid conditions.

Soil Testing

Soil analyses are made to determine whether soil is acid, alkaline, or neutral. The relative acidity or alkalinity of soil is measured by means of a pH scale. The figure 7 on the pH scale is neutral. Thus if we have a neutral soil, one which is neither acid nor alkaline, we say its pH is 7. If after testing soil we come up with a higher figure than 7 we say the soil is alkaline, and if the figure is lower than 7 we say the soil is acid.

If we have a limestone soil we can be certain that it is alkaline. Use of composts, mulches, peat moss, and cow or sheep manure help to overcome soil alkalinity. As organic matter in limestone soil is increased, individual rock particles become coated with algae, fungi, and other organisms. Thus the soil becomes less highly alkaline and approaches neutral on the pH scale. If the soil taken from the floor of a south Florida hammock is tested it generally will be found to be on the acid side of the pH scale. If a soil is highly alkaline, as in cases where a building lot has recently been cleared and where there is a great deal of free lime in the soil, liberal use of sulphur along with fertilizer and organic materials will help to reduce the alkalinity. If one is trying to grow acid-loving plants in limestone soil, such as ixoras or azaleas, application of aluminum sulphate may be required to obtain the degree of acidity the plants require.

Once a limestone soil has mellowed, following the accumulation of organic matter over a period of a few years, it becomes adaptable to a wide number of plants which would not thrive in raw limestone. On the other hand, in flatwoods areas, where the soil was originally acid, the accumulation of humus may increase the acidity and thus some minor correction through the use of lime may be necessary. Most ornamental plants, however, prefer soil which is slightly acid. This should always be kept in mind when the question comes up about the use of lime. Over-use of lime can cause serious soil problems.

Where ever one makes a garden, its quality in the end will be determined by the amount of effort put into the job. The only way to become an experienced gardener is to work with plants and soils, learning as much by the mistakes one makes as from the satisfying successes which one's efforts sometimes bring.

Use of Fertilizers

It doesn't take much experience in Florida gardening to learn that plants cannot be grown very well without fertilizer. Plants require nitrogen and minerals in quantities which the soil is not able to provide. Fertilizer is used to supplement plant needs.

Fertilizers are mixed today in an almost endless number of analyses. To understand their chemistry requires long years of schooling. This is not necessary for the gardener who needs mainly to know that his plants do require periodic applications of a balanced fertilizer. But it is helpful to know something about the mechanical action of fertilizer and the response of plants to it.

The terms "general purpose fertilizer" or "balanced fertilizer" are used to designate types which can be applied to most garden plants, including grass, shrubs, and trees. Such mixtures have relatively equal percentages of nitrogen, phosphorus, and potash. A very popular mixture is 6-6-6. These figures represent the percentages of nitrogen, phosphorus, and potash, always, in the order listed. A 4-7-5 mixture means that it contains four per cent nitrogen, seven per cent phosphorus, and five per cent potash. But there can be a difference in quality even though the analyses may be the same.

The higher the organic content of fertilizer the higher the price will be. The lowest-cost fertilizer will be one which is completely lacking in organic matter. An example would be a mixture whose nitrogen is derived from ammonium sulphate or ammonium nitrate. A fertilizer whose nitrogen is derived from such organic sources as sludge, tankage, tobacco stems, in addition to ammonium sulphate and ammonium nitrate, can be expected to be more costly. Organic sources may make up thirty to fifty per cent of the nitrogen analysis in a mixture. The higher the organic content the higher the cost is likely to be. Ammonium sulphate and ammonium nitrate, called chemical nitrogen, are readily soluble in water, and because of this they are quickly washed through the upper soil level by heavy rainfall or irrigation. Organic nitrogen, on the other hand, is not readily soluble, but breaks down slowly over a period of weeks or months, releasing small amounts of nitrogen throughout this time. Neither phosphorus nor potash is lost through leaching as rapidly as nitrogen is lost. In recent years a synthetic organic nitrogen has come into wide use. It is referred to by chemists as ureaform nitrogen. Like natural organic nitrogen it breaks down slowly in the soil, being released in small amounts over several weeks.

In addition to the dry type of fertilizer, usually sold in 25-, 50-, and 100-pound bags, two other kinds have come into use since World War II. They are soluble and liquid, highly concentrated, and usually sold in small containers. The soluble type is dissolved in water before being applied. The liquid form is diluted in water before being applied. Both types are widely used for fertilizing orchids, house plants, and even for lawns by running them through a sprinkler system. Both are more expensive than the bulk type. They leach rapidly in the soil and must be replaced frequently. Their convenience, however, is a major selling point and they have come into wide use. When applied according to directions they do not burn plants, and are thus usually much safer to use than the bulk form.

How to Apply Fertilizer

It is difficult to give directions for applying fertilizer, except in a very general way. Analyses vary from 4-7-5 to 12-12-12 and 20-20-20. It is a common practice to scatter fertilizer on top of the soil and wash it into the ground through irrigation. Only when preparing the soil for planting grass, shrubs, or trees is fertilizer incorporated with the soil. When fertilizing trees and shrubs on shallow, limestone soil it is necessary to spread it out over an area covered by the canopy of the tree's foliage. In limestone soil it is seldom possible for a tree to develop a tap root. Consequently the roots spread out within the top six or eight inches of the soil in a pancake of roots, almost perfectly flat on the bottom, and spreading out for a considerable distance. The fertilizer should reach to the limits of the tree's root system.

Few veteran gardeners try to measure the exact amount of fertilizer applied to a shrub or a tree. The material is scattered evenly over the soil, either by hand or with the aid of a fertilizer distributor. A thirty-foot tall mango tree might receive twenty-five pounds of a 6-6-6 fertilizer at a single application. This is roughly a fourth of a hundred-pound bag. The tree may receive three applications a year, or a total of seventy-five pounds. A shade tree which has reached a desired size may not require any fertilizer, if it is in good condition and does not produce heavy crops of flowers or fruits. After a flowering tree or a fruit tree produces a heavy crop it should be entitled to a fairly heavy application of fertilizer to assist in building it up for the next crop. Young trees are fertilized lightly and often, as we attempt to push their growth.

After the desired size is attained, we may need to fertilize only enough to maintain plants in good condition. Many a gardener makes the serious mistake of fertilizing grass, shrubbery, and trees on schedule three or four times a year, applying large amounts of plant food whether it is needed or not. This results in heavy growth, and consequently drastic pruning must be done.

In general, a 6-6-6 fertilizer is applied to grass at the rate of about twenty to twenty-five pounds per thousand square feet. A light application would be half this amount. In the vicinity of trees or shrubs, the application should be increased, of course. Beneath trees or shrubs, the application can be doubled. A small tree covering a 100-square foot area, or an area ten by ten feet, may receive about five pounds.

Fertilizer should never be applied close to the trunk of a tree or a shrub. Severe burning of the trunk as well as the larger roots may result. Where small trees or shrubs have been newly planted, there is little need to make large applications of fertilizer over a wide area about each plant. Roots are confined to a rather small area as yet, and applications should be fairly light but made at short intervals of one to two months. As a plant becomes established, fertilizer can be spread a little farther out as the roots spread out, and eventually applications can be increased in amount and reduced in number to two or three a year. Further information about fertilizing can be found in chapters dealing with lawn grasses and fruit trees.

Irrigation Necessary

It is necessary to provide some method of irrigation for every garden. Irrigation systems vary in extent from a couple of faucets to which a garden hose can be attached, to elaborate systems which are cut on automatically when the moisture content of the soil declines to a certain level. Many automatic systems are timed to cut on early in the day, in order that irrigation can be completed before the owner sits down to breakfast. Such systems make life easier for the gardener, but they are not necessary. It is necessary, however, to have plenty of water available if we are to grow a wide number of different kinds of plants.

Our ornamentals come from many parts of the world, where the rainfall may vary from 15 to 20 inches a year to as much as 150 inches a year. The rainfall may be evenly distributed throughout the year or it may all come during a rainy season of five or six months, leaving the rest of the year with little or no rainfall. Rainfall patterns and their effects on gardening are discussed in the chapter on *Climate and the Gardener.*

It is helpful to know a plant's origin and the kind of climate to which it is adapted. But this is hardly possible for the gardener to know except perhaps in a few instances. Furthermore, gardens are set up so that all plants are treated relatively the same way, and the gardener, usually with limited horticultural knowledge, naturally expects plants to react the same way. Plants, however, are not like this. They refuse to perform as the gardener may expect them to perform. Plants from areas of low rainfall may not survive under constant moist conditions. Flowering trees from areas of the world where there are distinct wet and dry seasons may not flower if they receive almost constant irrigation and the soil is never permitted to become dry. This is true of many of the cassias and acacias. The mango may not fruit well if it receives a constant source of moisture. Few mangos are grown in the tropics where there is heavy rainfall throughout the year. The better mangos are produced in tropical and semi-tropical areas where there are marked dry and wet seasons.

Wherever possible, plants should receive moisture conditions to which they are best adapted. Cacti, aloes, kalanchoes, and the euphorbias, such as the colorful crown-of-thorns *(Euphorbia milii)* should be planted in a well drained location, such as in an elevated boulder garden, and should be protected from over-watering. It is, however, impossible to protect such plants from the long periods of heavy rainfall south Florida experiences during the rainy season.

Thus the plants must be grown in well drained, well aerated soil. But dry-land plants do require moisture. In deserts where it never rains more than perhaps a trace, as along the coast of Peru, no plant life is seen except along water courses. Thus, during periods of long drought, desert type plants should receive an occasional irrigation. The irrigations should be done thoroughly, with the soil moistened to a foot or more. No additional irrigation may be needed for several weeks.

Very often it is not over-watering which injures plants, but the frequency of watering which does the greater harm. Where sprinkler systems are run for a short time each day, keeping the top inch or so of the soil moist, two types of injury are likely to result. First, the roots of plants tend to develop close to the surface. Secondly, perfect conditions are set up for several kinds of fungi to become established. Fungi attack the leaves of grass and ornamentals, as well as the roots of many plants. Over-watering may encourage the development of fungi in the roots of citrus, avocados, and hibiscus, often resulting in their deaths.

Until a plant is established it requires frequent watering, because it does not have an extensive root system to absorb moisture. A newly set out shrub or tree may require water every other day for a couple of weeks. How often will depend to a large extent on the type of soil and to the time of planting. A plant should be watered more often during dry weather, especially if there is lots of wind, than during humid weather. On the other hand, many a young plant has been lost because the owner neglected to water frequently enough during the rainy esason. Sometimes a shower may give the impression that the garden is well watered, but if one scratches beneath the moist surface one may find dry soil beneath. A newly set out plant should be watered on schedule except when at least one inch of rain has fallen.

For well established plants—plants which have been in the soil for at least a year—frequent irrigation is not necessary. If most shrubs and trees receive a thorough irrigation once a month during the dry season they should never suffer. Once-a-week watering is usually enough for grass, if an inch of water is put down when irrigation is done.

Water is taken up by plant roots. Inexperienced gardeners may spray their plants with a garden hose nozzle and believe they can do a good job of watering. But plants absorb little moisture through their leaves. They are not helped much by irrigation unless the soil is moistened to the depth where roots exist.

Much can be learned about plants by observing them during dry periods. Many plants wilt when the soil becomes dry. Other plants may shed their leaves. Plants in new growth may be benefited by irrigation during dry periods. Fruit trees may require periodic irrigation after the setting of fruit in order for the crop to develop. Ornamentals such as bougainvilleas and cassias, which flower best after going through a prolonged dry period, apparently are helped by irrigation after flowers begin to open.

Moisture can be conserved in the soil by eliminating weeds. Mulching aids the soil to retain moisture by slowing down evaporation. Windbreaks also help to reduce the loss of moisture in plants through transpiration. Cold weather affects the moisture tremendously. When the thermometer falls below 40 degrees, the leaves of plants may lose their moisture faster than they can absorb it from the soil. Thus injury may be suffered by the leaves of plants and grasses. For this reason it is a good idea to irrigate the garden thoroughly when a cold spell is reported on the way.

Preparing for Frost

Frost is a common winter visitor to south Florida, except for a very narrow strip along the coastline of the southernmost part of the peninsula and in the Florida keys. Hardly a winter passes that frost doesn't invade the southern part of the peninsula, covering most of the Everglades and reaching down as far as the outskirts of Homestead. The temperature may drop to freezing in open, unprotected areas on the outskirts of Miami. At least once in a decade freezing weather pushes its way to the shore of Biscayne Bay, and tender tropical plants and lawn grasses are brown for weeks thereafter.

The gardener should know something about the weather characteristics in the area where he lives. As a rule, the closer one gets to the coastline the warmer it is during cold spells. For instance, when the temperature drops to 30 degrees on the outskirts of Miami the temperature along Biscayne Bay may be 34 to 36 degrees. When the temperature drops to 30 degrees, many tender tropical plants may be injured or killed, especially if the low temperature persists for several hours.

It is coldest near ground level. A small plant with no protection between it and the sky is certain to be at the mercy of frost on a cold night. A tree with spreading branches may suffer no injury from frost, except during hard freezes. The temperature is always higher beneath the canopy of a tree than in an open lawn on a cold night. The most tender foliage plants grown on the floor of a jungle may escape injury from cold even on nights when the temperature drops to below 30 degrees in open areas. Under the protective canopy of trees the temperature may not drop below 38. Gardens in which there are several large trees are not likely to suffer much injury from cold except when south Florida is visited by a prolonged freeze, such as the area had in 1917, 1934, 1940, and 1958. Even then, tender plants growing under the protection of large trees may come through with little injury. If a freeze is accompanied by wind, severe damage may be done to plants even though they are under a canopy. The damage will be lessened, however, if a garden is protected by hedges or other screening on the northwest side. Keeping the soil moist also increases the protection.

By keeping up with weather bureau forecasts, it is possible to know one or two days in advance when cold weather is likely to strike south Florida. Thus the gardener is given plenty of time to provide protection.

How to Protect Plants

There are several ways to protect gardens and plants from cold damage. Small plants can be covered. Grove owners use bean hampers to cover year-old avocado trees. A young mango tree may be protected by covering its trunk with a bundle of grass or with any other kind of insulating material. The young mango tree is especially susceptible to freeze damage and likely to be killed by freezing weather until the tree has developed a canopy several feet across.

Rapid loss of moisture during cold weather is one of the main causes of lawns turning brown. If a lawn has been thoroughly watered the day before a cold wave is scheduled to move in, it will have a better chance of pulling through with small damage than if the soil is excessively dry. We don't have to worry about the ground freezing in south Florida, so we can safely water at the approach of cold weather.

Hungry plants are more susceptible to cold damage than robust plants. This is one good reason for making a fall fertilizer application, after the rainy season ends around the first of November. Ordinarily plants and grasses will be low in minerals and nitrogen at this time, after several weeks of leaching rainfall. A light to medium application of a balanced fertilizer will make up the losses of nitrogen and minerals and prepare plants for cool weather.

Heavy applications of fertilizer normally should not be made in the fall, as this may encourage new growth. However, it is not always possible to keep plants dormant throughout the winter, because if the temperature remains fairly high and tropical plants receive moisture they are likely to put out new growth. When in new growth plants are more susceptible to cold damage than when they are in mature foliage.

Except during hard freezes, it is entirely safe to run sprinklers over the lawn and shrubbery early in the morning to wash frost off the foliage. This will tend to prevent injury, which is almost certain to follow if the frost persists until the sun comes up. During hard freezes the water put out by sprinklers may turn into ice. This is especially likely where one uses a hand-held nozzle, and goes about spraying a mist onto the grass and shrubbery. Where one has a heavy sprinkler system, capable of turning out large amounts of water, there is little danger of the water turning into ice.

All weeds and mulch should be cleared from around small plants before a freeze. Mulch beneath larger trees which have formed a canopy does not matter. But there is good reason for removing mulch from around small plants. During the day the earth absorbs heat from the sun and releases it during the night. Weeds and mulch absorb little heat and consequently have little to give off at night. Freshly cultivated plants, as in a vegetable garden, are more likely to be injured by frost than where the soil has a crusty surface. A crusty surface can be quickly formed by irrigation.

It is good to have an accurate thermometer on hand to check the temperature which drops rapidly during a cold spell as soon as the sun sets. By checking the thermometer it is possible to get some idea about what temperature we are likely to have during the night. Here is a rule of thumb which might be remembered: If by 7 p.m. the temperature has dropped to 49 degrees, the chances are that it will drop to the vicinity of 32 degrees by sunrise. A temperature of 48 degrees at 7 p.m. indicates that at sunrise it may have 31 degrees, and so on. The temperature, however, can be influenced by several things. A light breeze may come up after midnight to halt the plunge of the thermometer. If clouds form they will reduce radiation of heat from the earth and thus prevent frost.

A hundred kinds of tropical fruits can be grown in Florida

Tomatoes and other vegetables are planted in the fall

Chapter 4

PLANT PROPAGATION

The supply of plants may be increased in several ways, from seeds, by rooting cuttings, by airlayering, and by division.

Seeds vary in size from the dust-like particles removed from orchid seed pods, to the fruit of the coconut. The smaller the seed the more difficulty one is likely to have in growing it. Orchid seeds are planted in a special medium, in sterile flasks. Seeds of many annual flowers are so fine that once they are scattered in the soil they cannot be distinguished from the dark particles of humus. Small seeds are best planted in flats or in pots of sterile soil.

Small amounts of soil can be sterilized in the oven of a stove, by putting it in a large container and adding a small amount of water to create steam. Quick results are obtained if the pan is covered. About twenty minutes of steaming should be enough to kill all microorganisms, soil insects, and weed seeds.

Larger seeds, from palms or ornamental shrubs, may be planted in flats of soil, made by combining perlite or crushed granite with a small amount of well pulverized and screened sphagnum moss. When individual seeds sprout they can be lifted and placed in containers of good soil. The original planting medium contains no nutrients and if plants are left temporarily in such mixtures as perlite and sphagnum, they should be fed with liquid or soluble fertilizer. Young seedlings should be grown in as much light as they can take. If grown in deep shade they become leggy and weak. Most young plants thrive best when grown under slats or plastic which provide about forty per cent shade. In the fall and winter, when annuals and vegetables are planted in seed flats for later transplanting to garden rows or beds, they should be grown in full sun from the beginning.

A great many ornamental plants may be propagated from cuttings, either mature wood cuttings or from tip cuttings which are selected just before new growth reaches maturity. There are few plants which cannot be propagated if the cuttings are kept under mist. Nurserymen use automatic systems for cutting on mist at regular intervals over cutting beds. Root-inducing hormones aid in the rooting of difficult plants. Cuttings must be placed in a medium which is well drained, yet which retains sufficient moisture to prevent the cuttings from drying out between waterings. A mixture of perlite and either sphagnum or peat moss makes a suitable medium. It may be helpful to cover cutting flats with transparent plastic in order to aid in the holding of moisture. If plastic is used it is usually possible to root cuttings on which a few leaves have been retained. This hastens the rooting process.

How to Airlayer

Airlayering is usually the best system for the homeowner to use for propagating plants, for two reasons. First, it is easy to do once the method has been learned, and, secondly, fairly large plants can be so propagated. A branch of suitable size and shape is selected, preferably one whose foliage receives sun most of the day. The beginner will do well to start out with rather small branches, the size of a pencil or a little bigger. At a point on the branch where you wish to induce roots to grow the bark is ringed in two places, about an inch and a half apart. The bark between the ringed areas is now removed, down to the wood. No part of the soft inner bark must be left. This can be scraped off the wood with the knife blade, if any of it persists after the outer bark is taken off. This soft inner bark is the cambium layer, and if any is left in the girdled area a bridge will be formed over the gap and rooting may not take place. Next, take a handful of moist sphagnum from a bucket of water and squeeze out enough of the water so that the sphagnum no longer drips, but is still quite wet. Distribute this around the girdled part of the branch, and cover with a sheet of aluminum foil, large enough for over-lapping, then twist both ends around the branch.

Roots should begin to form about the girdled area in six or eight weeks. After the sphagnum is full of roots, the branch can be cut just below the aluminum foil—between the aluminum foil and the trunk of the tree—with the aid of sharp clippers. Now very carefully remove the aluminum foil and set out the new plant in a container of rich soil. Extreme care must be exercised to prevent injury to the roots. Soil should be washed into the container and about the roots with the aid of a garden hose.

Keep the young plant in part shade for several days, until it becomes established, and then carefully remove it to full sun. Under normal conditions it should be ready to plant out in a permanent location within a couple of months. Again care must be exercised when removing the plant from its container which by now should be full of roots. If handled roughly, not only will many of the roots be damaged, but there is danger of breaking the entire root system from the plant stem.

Most plants can be propagated by the airlayering method. The best month to induce rooting of airlayered branches probably is in May, June, or July, but many gardeners use this method of propagation at all times of the year. One of the main reasons for lack of success with airlayering is the failure to do a good job of girdling the branch before the moss is put on. If, after six weeks no roots have formed, the girdled area should be examined to see if it has calloused over. If so, the girdling should be done over and fresh sphagnum put on.

Propagation by Division

Many plants are propagated by division, such as the banana, bamboo, the rhapis cluster palm, and most bulb plants. Banana plants are usually increased by removing suckers from the base of an old tree or old clump of trees. Bamboo is usually propagated by the removal of a small portion of a larger clump. Sometimes three or four large bamboo stalks, or culms, are included in the portion of roots and underground stems removed. The tall bamboo stems are cut back to within three or four feet of the ground and this section is carefully dug from the old clump and transplanted. This method of propagating bamboo is by no means easy. The roots and underground stems are very tough, and both an ax and a good grubbing hoe, as well as strong muscles, are needed for the operation.

The rhapis palm also is propagated by the removal of three or four stalks along with their underground stems and roots, from an older clump. These new plants will be very weak, however, and should be kept in containers until they are well established before being planted in a permanent location. This may take six months. Gingers and heliconias are propagated from division, but ordinarily this method of propagation is not recommended for the strelitzia. Division of the strelitzia can be done only with great difficulty. It is better to buy vigorous young plants from a nurseryman rather than to mutilate a fine old clump in order to get a few off-shoots.

Propagation by division is best done in the early part of the growing season, except in the case of such bulbs as the daylily which flower in late spring. Normally both the daylily and the amaryllis are divided in August. These plants need not be divided, however, unless they have been growing in the same bed for several years and are very dense.

Budding and Grafting

Fruit trees and outstanding flowering trees are often propagated by budding or grafting. Many plants may not thrive on their own roots. For instance, the gardenia is nearly always budded or grafted onto the root stock of an African species of gardenia before it is planted outdoors in south Florida. The cultivated gardenia requires an acid soil and does not do well on its own roots in limestone soil. It also is highly susceptible to nematode damage when grown on its own roots. The African species *(Gardenia thunbergia)* is affected little by either limestone or nematodes.

Although the sweet orange may be propagated by airlayering, it does not thrive under most conditions on its own roots. In limestone soil it grow best on Cleopatra tangerine stock, while in low, moist which is subject to flooding, it does best on sour orange root stock.

The mango may be propagated by airlayering, which is the method used in India, but few homeowners have enough background in horticulture,—and patience—to propagate mangos by this method. Usually budded trees are purchased from nurserymen. The lychee is easy to propagate by airlayering, while the carambola and the sapodilla are somewhat more difficult.

Numerous methods of budding and grafting are used for the propagation of mangos, avocados, citrus, and other trees. These methods are covered by several texts, including bulletins published by the Agricultural Experiment Station at Gainesville. These may by consulted in libraries or at county agents' offices. However, methods of budding and grafting are usually learned through practical demonstration rather than from the reading of texts. Beginners interested in learning this art should start with plants which are easy to graft, such as hibiscus and the loquat. The budding of the mango is much more difficult, and the budding of a sapodilla is a job for the expert horticulturist.

The quality of your Florida garden will depend more upon your own efforts than upon the natural growth of plants in a favorable climate.

Shrubby plants must be carefully selected for each location

A shrub's tolerance of pruning is an important consideration

Chapter 5

SHRUBS AND THEIR USES

The shrub is the most important plant in any landscape plan. Trees, palms, vines, ground covers, and annuals are of secondary importance. Ordinarily a small garden can support two or three trees at the most, unless the trees are very small. Only a few palms, such as *Rhapis excelsa* and *Chrysalidocarpus lutescens,* are suitable for screening or for background planting. One can do without annuals or roses or ground covers in a garden, but one could hardly do without shrubs.

There are many kinds of shrubs, in many forms, textures, and colors. They vary greatly in growth habit. Many become small trees if left unpruned. Others are low and sprawling, and may be listed also under ground covers. Some are viney in growth habit, such as the jasminums, and may be listed under vines as well as under shrubs and ground covers. Even certain foliage plants may be listed under shrubbery, including the monstera, banana, dracaena, and alpinia. The suckering palm *Rhapis excelsa,* and also the pandanus, are used as shrubs. Several fruit trees may be included, such as the cattley guava, one of the most useful of background shrubs or small trees, and the Barbados cherry, which is as colorful when in fruit as many flowering trees.

Landscaping can be done by using no other plants except shrubs and grass. Shrubs serve many purposes; for screens and hedges, for foundation plantings, and for accent, for facing taller shrubs, for ground covers, and for color or unusual form. They should be chosen carefully. When several kinds are planted together they should be selected for agreeable form, texture, and color. Marked incongruity in shrubbery plantings can give the impression of a weedy garden. Some knowledge about the growth of each shrub is necessary.

Large shrubs, used for background or to screen objectionable views, should be planted some distance from the property line so that their branches will not extend over on a neighbor's lot. A cattley guava will reach eight to ten feet high, unless severely pruned, and it will develop a spread of close to eight feet. Thus it should be planted a minimum of four feet from the property line. It should be remembered that zoning laws in some municipalities prohibit hedges above five feet, but three or four cattley guavas, set in a group to be used as a screen against a particularly objectionable view or to muffle noise from a neighbor's air conditioning equipment, are usually permissible. Where there is not enough space for spreading shrubs, the common aralia *(Polyscias guilfoylei)* may be used. It grows upright, with little spread, and can be planted within eighteen inches of a lot line. In time, without trimming, the aralia becomes woody and ugly.

Use Appropriate Shrubs

Planting of large shrubs, such as hibiscus, ligustrum or ixora, beneath windows, so that they constantly must be cut back or rounded with pruning shears, should be avoided. There are many low growing plants which can be used for such locations, including plumbago, holly-leaf malpighia, jasminums, and the sprawling forms of carissa, which need little pruning.

Where formal, sheared hedges are desired the gardener has a choice of several species. The more common ones used for this purpose are the Surinam cherry *(Eugenia uniflora)*, *Murraya exotica, Ixora coccinea, Ligustrum japonicum, Ligustrum sinense,* and *Pittosporum tobira.* Where very tall clipped hedges are desired, *Ficus retusa* and *Ficus benjamina* may be used. Although it is not widely planted in Florida, the commercial olive *(Olea europaea)* makes a very good clipped hedge.

For the purpose of general description rather than exact botanical classification, shrubs might be divided into two groups, hardwoods and softwoods. Examples of hardwood shrubs are carissa, Surinam cherry, cocoplum, severinia, jacquinia, ligustrum, malpighia, murraya, cattley guava, and podocarpus. Softwood shrubs are acalypha, aralia, night and day blooming jasmines, clerodendron, crotons, duranta, crape jasmine, hibiscus, Turk's cap, dwarf poinciana, thryallis, and vitex.

Plants of the hardwood species are almost invariably superior in form and texture to plants of the softwood species. The hardwood species are slower growers, require less pruning, tolerate pruning better, are generally heartier and less subject to diseases and insect troubles than the softwood plants. As with everything, there are exceptions. Some of the most interesting plant forms and leaf textures are found in a softwood group, the aralias. *Aralia elegantissimum* is an example.

Because of their slow growth the hardwood species generally are more costly to buy than the softwood species. Nurserymen propagate few of them. *Eugenia coronata* is an example. However, in recent years nurserymen have learned a great deal about how to hurry the growth of plants and the difference in costs of shrubs is not so great as it used to be. Cost is based mainly on how long a plant must be kept in the nursery. The hardwood species, being more difficult to propagate, used to be grown mainly from seeds. Young seedlings were very slow to grow. Now most plants are propagated from cuttings under mist.

Many Plants Difficult to Find

A list of tropical plants which could be used as shrubs would run into hundreds of species. Nurseries, however, propagate a relatively small number. The landscape architect, when planning a garden, must stick with plants which are available. The homeowner must do the same, unless he is willing to search for propagating material and grow his own. If he is interested enough he can find the species from which he might obtain airlayers or seeds, as for instance, the many very fine native plants which grow in local hammocks.

To use plants well in a landscape requires a great deal of knowledge about the plants themselves. One must know the ultimate size and the rapidity of growth of a plant and whether it takes pruning well. And, of course, each plant has its own character and qualities which one can get to know only from long association and observation. It is next to impossible to recommend, in every instance, the exact plants to be used in a given situation. Much depends on the sun or wind exposure, the type of architecture, the taste of the homeowner, and upon the relationship of other plants already selected.

Shrubs vary greatly in character, and the gardener should consider this when selecting them for planting. They might be classified according to bushiness, color of flowers or fruit, and texture, and color of foliage. It is obvious that there is a great deal of difference in character between a hibiscus and a carissa. The dwarf poinciana is very different in character from either of these. It might be possible without reference to exact botany, to group plants, in a simple descriptive way, in one of these three categories.

For instance, any shrub with color and leaf texture similar to that of the hibiscus might be listed in the "hibiscus character" group. Such plants might be the Turk's cap (malvaviscus), night-blooming and day-blooming jasmines, clerodendron, poinsettia, chenille plant, roselle, and datura. These are rapid growers, and their characters are so similar in a number of ways that they frequently can be used together in landscape design and planting.

In the carissa group, which includes many plants that may be slower in growth than those in the hibiscus group, we might include the ligustrum, Surinam cherry, citrus, members of the myrtle family such as the native eugenias, and cattley guava, several kinds of ixora and also the cocoplum. Their foliage is stiffer and more glossy than that of the hibiscus group. Color, form, and texture are so similar that they might be planted near one another without being inharmonious.

In the third group, the dwarf poinciana, we have a number of fairly straggly plants with either fine foliage or open foliage which sets them apart from other bushy members of the hibiscus or carissas. With the dwarf poinciana one might include the jumbie bean, coral plant, scrub acacia (*Acacia farnesiana*), parkinsonia, calliandra, and several cassia species.

Create Own Categories

The gardener can make up his own plant groups, including shrubs that seem to go together harmoniously, according to his sense of values. We have not listed, for instance, such plants as the acalypha or the croton. Nor have we discussed the varieties of bougainvillea which can be trained as shrubs. These are among the more brilliant plants, and their use offers still another problem. They have their places in the tropical garden, and deserve wide use, but many gardeners make the mistake of giving over their gardens entirely to gaudy plants. In general, the colorful plants should be reserved for planting after the basic shrubs and trees have become established.

Most certainly the gardener should not be so extreme as to make a choice from only one group and neglect the others, in order to develop a purist's garden in which there was no contrast in shrub form or unlikeness in texture. In fact, the best garden will have all categories in order to provide a never-ending variation. How far to go with this sort of thing—just exactly where to stop—is a problem that can be answered only by the gardener himself; and this will be determined by his experience and his taste.

But it does not follow that one who has had a considerable amount of experience as a gardener will have superior taste. For there are persons who have good taste in gardening although they have never planted a shrub or pulled a weed. These are observing people who have spent much time studying and enjoying gardens everywhere they go. Their sense of what is right in landscape design may be far superior to that of gardeners who have spent much of their lives growing and caring for plants.

When planning and planting a garden, it is good to look into the future, when the tiny plants you set out, wondering if they will ever get started, become so large that they must be constantly pruned to be kept under control. The landscape architect, when drawing up the plans for the garden of a wealthy homeowner, can use fully developed trees and shrubs. He can place background shrubs where they belong, an eight-foot cattley guava here, and several six-foot ligustrums there. In front of these he can set just the right kind of facing shrubs, of just the right size, and so on until he has the right effect. Such a plan is first drawn on paper, then specifications are prepared for use by the landscape contractor who does the planting. When the owner moves in, he has a landscaped garden as fine as though several years had been spent in developing it.

Not evervone can afford such a completely landscaped garden. Eight-foot plants, moved from nurseries by tree trucks, cost a great deal of money, and labor is expensive. But even the smallest cattley guava or the smallest ligustrum eventually gets to be six or eight feet tall. They represent an investment in patience. But if such plants are not set in the right place from the beginning, they can become problems as they reach maturity.

Suppose plants which eventually will become large shrubs are set where they have no freedom for growth, such as beneath windows, or in places where they spread out over walks or driveways. This means cutting back one or more times a year. They cannot develop a natural form, and they become the sheared and characterless specimens which are so numerous in south Florida, even in gardens where the owners can afford to do better. Thus, when planting a shrub it is a good idea to know what its ultimate size will be and to make allowances for full development of character.

Value of Windbreaks

As suggested elsewhere, only tough plants should be used for windbreaks or for background. These plants serve not only as background for your fine ornamentals, but they serve as a defense against cold winds. In older gardens the first planted shrubs become small trees. Two or more facer shrubs can be planted in front of these, in a sort of mother-child effect. This system, if used in an orderly way about one's garden could become monotonous, but the chances for such an orderly development of a one-two-three planting of tall, medium, and low shrubs is hardly likely. Furthermore, such plantings can be broken up by palms, trees, and shrubs of various shapes and leaf textures.

Every shrub should be chosen for a purpose, as well as for the effect which you wish to achieve. No shrub should be planted without first considering whether it will associate well with other plants in the landscape, and especially with near-by plants.

Below is a list of shrubs, as well as small trees which are commonly used as shrubs. Descriptions are not complete enough to make identifications. Plants should be seen before purchased rather than be selected from description. This information should be helpful when one is trying to find out the average size of plants and something of their characteristics. Methods of propagation also are indicated. Most of the plants listed here can be found by shopping among nurseries. The slower a plant is in growth the more difficult it is to find. But occasionally one finds a nurseryman who will have one or more of these very difficult plants, despite their slow growth. Those which the shopper cannot find, if he wants them, he will have to propagate himself.

ACALYPHA HISPITA. Chenille plant. Height five feet, spread four feet. Upright and dense. Rapid growing, softwood. Leaves large and bright green. Flowers tassel-like and bright red. Used for hedge and accent. Cuttings.

ACALYPHA WILKESIANA. Erroneously called red aralia or croton. Height five feet, spread four feet. Upright and dense. Rapid growing, softwood. Leaves large, red or bronze. Used for hedge, filler or accent. Cuttings.

ARALIA ELEGANTISSIMUM. Height to eight feet, making small tree if not controlled. Matúre leaves large, glossy green, with interesting pattern. Used as background shrub or several together for screening. Grows too tall to be planted near foundations, but is widely used for this purpose. Seeds, cuttings or airlayering.

ARALIA, COMMON. See Polycias guilfoylei.

ARDISIA. Several species of this hardwood shrub. *Ardisia crispa* is best known. Height five to seven feet. Spread four feet. Fast grower. Glossy green leaves. Flowers in white clusters. Fruit red, turning black. Used for screening, hedge or background shrub, in sun or shade. Seeds.

BAUHINIA GALPINI. Half-climbing, sprawling hardwood. Height six feet, spread eight to ten feet. Leaves medium green. Flowers red, in summer. Of use only where large sprawling plant desired. Can be trained on wall or fence. Airlayering or seeds.

BIXA ORELLANA. Height ten feet, spread eight feet. Rapid growing softwood, making a small tree. Leaves large, heart-shaped. Flowers rose. Fruits spiny, bright red. Of use where large accent plant desired. Seeds.

BRUNFELSIA AMERICANA. Height five feet, spread three to four feet. Spreading, open shrub. Leaves glossy green. Flowers white, turning pale yellow, borne in clusters. Does well in shade. Seeds.

BRUNFELSIA CALYCINA. Height four feet, spread five feet. Spreading, dense shrub. Flowers rich purple. Does well in shade. Seeds, cuttings or root sprouts.

CALLIANDRA. Several species, four to seven feet tall, with equal spread. Leaves fine. Flowers bright red or pale pink. Used as accent plant, but difficult to fit into landscape because of form and texture of small leaves. Seeds.

CARISSA GRANDIFLORA. Several varieties of this hardwood shrub, including such names as Boxwood Beauty, Linki, and Bonsai. Seedlings, very spiny, grow to eight feet, with five-foot spread. Named varieties are sprawling, two to four feet high. Leaves stiff and dark green. Fruit glossy red, edible. Sprawling varieties widely used by landscape architects as facer or filler plants, ground covers, foundation plants, and in planter boxes. Can be grown from seeds, but varieties can be maintained only if propagated by airlayering or by cuttings.

CESTRUM DIURNUM. Day-blooming jasmine. Height to eight feet, spread four feet. Rapid growing, brittle softwood, tending to be leggy. White flowers fragrant in daytime. Fruit black. Of use only as background plant, but little used because of openness. Seeds, cuttings or airlayering.

CESTRUM NOCTURNUM. Night-blooming jasmine. Height to eight feet, spread four feet. Rapid growing, brittle softwood resembling day-blooming jasmine. Greenish white flowers are very fragrant at night. Fruit white. Of little value as landscape plant, but widely used for fragrance of flowers. Seeds, cuttings or airlayering.

CHAMAEDOREA SPECIES. See Palms.

CHRYSALIDOCARPUS LUTESCENS. See Palms.

CHRYSOBALANUS ICACO. Cocoplum. Height to seven feet. Variable hardwood, slow from seeds but faster from cuttings and airlayers. One of outstanding shrubs propagated. Foliage glossy green. New leaves on some varieties reddish. Fruit black or pink. Pulp and kernel edible. Used by landscape architects for background, screening or as sheared hedge. Seeds, cuttings under mist, or airlayering.

CLERODENDRUM. Several species, three to six feet tall, often tending to become weedy in appearance and in aggressiveness. Grown for white or red flowers and also for unusual ornamental fruit. The clerodendrums are hardly suitable for permanent planting, being difficult to maintain in a desirable form. They are usually planted to provide quick color. Most of them thrive in full sun or part shade. Root sprouts, seeds, or cuttings.

COCCOLOBA DIVERSIFOLIA. Pigeon plum. This is a native which is a relative of the seagrape. It is a fairly rapid growing tree, but can be kept under control by pruning. Its large glossy leaves are produced in dense numbers and the small tree should be excellent for screening. Easily propagated by airlayering. Large branches may be taken.

COCCULUS LAURIFOLIUS. Here is another dense growing shrub, with slender shiny, very dark green leaves. Takes pruning well, and although it makes a tree in time if left unpruned, it can be kept at desired height. Airlayering.

CODIAEUM VARIEGATUM. Croton. There is an almost endless list of varieties of this colorful and popular shrub, four to six feet tall and up to four feet of spread. Color combinations of greens, yellows, reds, and browns. Varieties differ in tolerance to sun and shade; but most crotons do best in part shade. Crotons must be used with great care or they will steal the show and upset the best landscape plan with their often too brilliant colors. Airlayering or hardwood cuttings. Nurserymen propagate tip cuttings under mist.

CONOCARPUS ERECTUS. Buttonwood. Although this native of the salt flats and seashores of Florida and the Caribbean becomes a large tree, it is widely used for background or where high screening is desired. Two varieties are propagated by nurserymen, one with medium light green leaves and one with silvery leaves. Airlayering or cuttings.

DOMBEYA NATALENSIS. Rapid growing softwood shrub to eight feet, with almost equal spread. Large heart-shaped leaves, similar to hibiscus tree, with many clusters of pink flowers. Suitable only for temporary use. The dombeya is susceptible to nematodes, and even when grown well this shrubby tree becomes weedy in appearance. Seeds.

DOVYALIS HYBRID. Florida apricot. Abyssinian gooseberry. This desirable shrubby tree will reach twelve feet in height, with equal spread, and takes pruning well. Light green in foliage and suitable for screening. Produces excellent fruit for eating out of hand. Propagated by airlayering or seeds.

DURANTA REPENS. Golden dewdrop. Willowy, sprawling shrub to eight feet, with equal spread. Panicles of lilac-colored flowers are followed by bright orange fruits, from which plant derives its common name. Because of its open branches and small foliage, this plant is hardly suitable for use as a screen, and it seems to go well with few other common shrubs. It is best used as a large, sprawling shrub where such a form is desirable. It is not suitable for a small place. Seeds or airlayering.

EHRETIA MICROPHYLLA. Tea plant. False tea. Philippine tea. An irregular shrub which is easily kept in place by pruning. Especially suitable for shady or semi-shady locations, but grows best in full sun. Small leaves are a shiny deep green. Flowers and fruits are insignificant. This plant is seldom seen in nurseries any more, but used to be quite popular. Seeds.

ERVATAMIA CORONARIA. Crape jasmine. Tabernaemontana. A free flowering shrub which is often confused with the gardenia. Grows to four or five feet, with equal spread when properly trained. The main objection to this shrub is that it tends to drop its lower leaves. Its large glossy green leaves are enough to recommend this plant, but it also produces an abundance of white, waxy flowers up to nearly two inches in diameter. Hardly suitable for screening or for use where dense shrubbery mass effects are sought. Thrives in shade and on north side of house. Airlayering or cuttings.

EUGENIA BRASILIENSIS. *Eugenia dombeya*. Grumichama. A hardy evergreen shrubby tree, but very slow in growth. Produced from seeds taken from dark black edible fruits.

EUGENIA BUXIFOLIA. This has been a favorite native plant used for years by the Dade County Park Department for screening and background planting. Small, glossy green foliage is dense. Tree takes pruning well and can be kept at any size desired. Growth slow in early stage, but once established, growth is fairly rapid. Like most of the other eugenias, this one is seldom found in nurseries. Seeds.

EUGENIA CORONATA. Another desirable background shrub with deep green foliage, but very slow to grow. Like *Eugenia brasiliensis,* this one should not be planted where a quick effect is required but should be set in a place where you can afford to wait. It might be used in front of a planting of aralias. The aralias may be removed after the better shrub has matured. Seeds.

EUGENIA UNIFLORA. Florida cherry. Surinam cherry. A desirable fruit producing shrub growing to twelve feet or more with eight-foot spread. Lends self to pruning and can be maintained as a clipped hedge if desired. Thus shrub is best used where rather large background shrubbery masses or screens are required. Foliage glossy green. Fruit crimson. Also covers itself in spring with creamy white flowers. Seeds.

EUPHORBIA PULCHERRIMA. Poinsettia. Height to six feet, spread about five, with form depending on correct pruning. This is one of the first plants tried by the newcomer to Florida. For a permanent position in the landscape, the poinsettia has its limitations. It does not fit well with many other shrubs, and is such a rapid grower that during much of the year it resembles a weed. Instead of incorporating the poinsettia as a permanent plant in the landscape, it might be better to grow it in containers and bring it out for display while in bloom at Christmas time. Easy to propagate from mature cuttings taken in late winter or early spring.

FORTUNELLA JAPONICA. Kumquat. Height to eight or ten feet, spread to four or five feet. This citrus is a slender growing tree with deep green foliage, and is very decorative when filled with orange-yellow fruit. Best used where an accent plant is needed. May be propagated from seed, but budded trees ar more dependable.

GARDENIA JASMINOIDES. Height to eight feet, with almost equal spread, when grown as a grafted plant. Because it is subject to nematode attack and also requires an acid soil, the gardenia should never be planted on its own roots in Florida. Nurserymen graft it onto the root system of a wild African gardenia. Several varieties are available. When used well the gardenia can be a striking shrub in the landscape, but should be planted so that it has plenty of room for growth. May be propagated from cuttings, but such plants can be grown only in peat moss.

HIBISCUS. Height to twelve feet, spread to eight feet, but most varieties are smaller. Over 5,000 hibiscus hybrids have been developed in Hawaii, and well over 500 varieties have been listed by the American Hibiscus Society as being cultivated in Florida. But only a few hibiscus varieties are used in permanent landscape planting. So many of the hybrids produce striking individual flowers, but the plants may be weak and thin of foliage. Only the proven, vigorous varieties should be used for basic planting, such as for hedges, screening or for background. Gardeners especially interested in hibiscus should set aside a special area for a collection of varieties. Nurserymen who specialize in supplying plants for landscape architects and landscape contractors usually carry the more vigorous kinds of hibiscus, which have been selected because of their special vigor and lasting ability. The LaFrance is an example of a variety which was selected because of its vigor and resistance to the multitude of troubles which befall this group of plants. The more vigorous hibiscus may be propagated from cuttings or by airlayering. Many of the weaker hybrids must be grafted onto the rootstock of vigorous varieties, such as the common red hibiscus, as they do poorly on their own roots.

IXORA. Several species and varieties of this colorful tropical shrub are in wide cultivation in south Florida. Most common is *Ixora coccinea,* an almost ever-flowering species which is commonly used for sheared hedges. It is one of the few plants which flowers well when maintained under sheared conditions. More popular as an individual shrub is the giant flowering type, *Ixora macrothyrsa,* which produces flower clusters eight inches across. Several varieties are on the market, their flowers ranging in color from yellow to pink, orange, orange-scarlet, and scarlet. Plants reach eight feet or more in height, when permitted to grow freely, with a spread about equal to the height. Another species, *Ixora parviflora,* makes a small tree, but can be trained as a sprawling shrub. This is a very desirable plant when well grown, because of its glossy, deep green leaves. It will thrive in part shade. All ixoras are propagated from cuttings or by airlayering.

JAQUINIA ARMILLARIS. Height to seven feet with five foot spread. Leaves gray green, fruit scarlet. An excellent shrub for planting near the seashore, but very slow in growth. Seeds.

JASMINUM AMPLEXICAULE. *Jasminum ilicifolium. Jasminum undulatum.* A strong growing viney shrub that is similar in growing habit to *Jasminum gracile* and is frequently used by landscape architects. The pointed, glossy green leaves reach four inches in

length. The white flowers, over one inch in diameter, burst out of pink tinted buds, and are in evidence most of the year. Cuttings. ing vine is grown mainly as a shrub or ground cover, and is desirable for its evergreen, lustrous foliage and its freedom from
sirable for its evergreen, lustrous foliage and its freedom from
pests. It thrives in full sun or in part shade, and is a fairly rapid grower. Clusters of small, white flowers are produced at intervals. Cuttings.

JASMINUM PUBESCENS. Star jasmine. A sprawling or climbing plant widely grown as a low shrub. Successful both for ground cover or for sprawling, facer type of shrub. Recommended for outdoor planting boxes. Grows well in full sun or part shade. Produces clusters of white, star-shaped flowers most of the year. Cuttings.

JATROPHA HASTATA. Height to five feet, open in form and
outdoor planting boxes. Grows well in full sun or part shade. Pro-
duced over long periods. Seeds or cuttings.

JATROPHA MULTIFIDA. Height to fifteen feet with almost equal spread. Rapid growing shrub with leaves nearly a foot across. Flowers scarlet. Should be considered a temporary plant in the landscape. It quickly becomes weedy. Seeds.

LEUCOPHYLLUM FRUTESCENS. *Leucophyllum texanum.* Texas silver-leaf. Sprawling, dense shrub to six feet, with small silvery leaves. Rose-violet flowers are produced several times a year. Thrives best in full sun. Slow grower and for this reason only a few nurserymen carry. Prices usually are high. This very fine shrub, well adapted to south Florida, is a notable exception to the rule that semi-desert plants do not thrive in a moist climate. It is native to the Southwest and Mexico. Seeds. Nurserymen often order small plants from Texas.

LIGUSTRUM JAPONICUM. *Ligustrum lucidum.* Wax privet. Height to fifteen feet, spread to eight or ten feet. Strong growing, hardy, excellent background shrub or small tree. More often grown as hedge plant, sometimes sheared. Leaves deep green and waxy. Growth compact. When grown as a tree it is covered with clusters of creamy white flowers once a year. Cuttings or airlayering.

LIGUSTRUM SINENSE. Chinese privet. Height to twelve feet, with growth more upright than *Ligustrum japonicum.* Leaves glossy green, but smaller than leaves of *L. japonicum.* Foliage dense. Takes clipping well. Cutting or airlayering.

MALPIGHIA COCCIGERA. Holly-leaf malpighia. Height to four feet, with equal spread. Some varieties, such as Low Boy, are much lower, with height to two feet and spread to four feet. The malpighia is one of the better shrubs for south Florida. Low-growing varieties are widely used for ground covers. Leaves small and holly-like in form. Persistent bloomer, producing pink flowers, followed by a small bright red fruit. Seeds, airlayering, and cuttings under mist.

MALPIGHIA GLABRA. Barbados cherry. West Indian cherry. Acerola. Height to twelve feet, spread to ten. This shrub, or small tree, produces several crops of bright red, vitamin-loaded fruit in spring and summer. Can be used as a background shrub or as a hedge, but if sheared its fruiting will be reduced. It is best grown as a large round-headed shrub. Seeds, airlayering, or cuttings under mist.

MALPIGHIA SPECIES. This is a species of malpighia introduced several years ago by the Fairchild Tropical Garden. It is a vigorous grower but somewhat smaller in every way than *Malpighia glabra.* Its abundance of small red fruits make it highly ornamental. There is even a smaller form of the *Malpighia glabra.* It gets to be only about three feet high and is sold as a dwarf variety of *Malpighia glabra.* Seeds or airlayering.

MALVAVISCUS GRANDIFLORUS. Sleeping hibiscus. Turk's cap. Height to seven feet, with equal spread. Foliage light green. Grown chiefly for its pendulous flowers which never fully open. Growth is as vigorous as that of a weed. Should not be planted on a small place. Cuttings.

MURRAYA EXOTICA. Orange jasmine. Chalcas. Height to ten feet, spread to eight feet. Small tree suitable for background or screen, or may be used as sheared shrub or hedge. Leaves small and glossy light green, but produced densely on healthy plants. Masses of white flowers, followed by orange-red fruit. Seeds or airlayering.

NERIUM OLEANDER. Common oleander. Height to ten feet, spread to eight. Elongated, glossy green foliage. Grown mainly for flowers, of which there are several colors. White and pink are the most common seen. Chief objection is the plant's highly toxic character. Flowers or leaves eaten by cattle or children may cause death. Cuttings.

OCHROSIA ELLIPTICA. Kopsia. Upright or spreading shrub to eight feet. Leaves a bright glossy green. Fruit, the size and shape of that of a carissa, is bright scarlet. Outstanding background shrub or screen, but in early stage is slow grower. Seeds.

PANDANAS. Screw pine. Several species of this useful plant are propagated for landscape use in south Florida, but *Pandanus utilis* is most widely sought because its leaves are preferred for weaving. Screw pines reach fifteen feet high or more in height, becoming small trees in time, with bare lower trunk and branches. They are, however, often used where dense screening is desired, especially a screening which cannot be easily penetrated. Aerial roots develop on old specimens, reaching to the ground and serving as props. A picturesque effect results, and some of these very old plants make rather bizarre specimens. Seeds and branch cuttings.

PEDILANTHUS TITHYMALOIDES. A succulent type of plant, which although not a true shrub, is often grown in dense clumps as a shrub. Four or five feet in height, with deep green leaves and red flowers in terminal groups. Requires little care. Cuttings.

PENTAS LANCEOLATA. A most superior plant where quick shrubby growth and quick color is desired. Plant, however, is short-lived. Height to four feet with equal spread. Flowers in red, pink, lavender, and white. Terminal clusters should be removed as quickly as flowers fade to induce continuous flower production. Cuttings.

PITTOSPORUM TOBIRA. Outstanding landscape shrub. Takes pruning well, but must be kept clean of thrips and scale insects. May be used as a sprawling shrub, as a hedge or as a small tree. Both dark green and variegated varieties available. Popular plant in landscapes which have Oriental influence. Withstands salt spray. Cuttings or airlayering.

PLUMBAGO CAPENSIS. A sprawling plant with small light green leaves and blue or white flowers. Height to two or three feet, with equal spread. Thrives in full sun, although it does grow in some shade. Might be grown as ground cover where blue flowers are desired. Cuttings.

PLUMERIA RUBRA. Frangipani. Spreading small tree or large shrub to fifteen feet or more. Should be grown as tree, except where special effects desired. Because leaves are subject to fungus this tree is without foliage a large part of the year. Therefore care must be exercised in its use. Grown mainly for large clusters of pink, yellow or white flowers. Cuttings.

POINCIANA PULCHERRIMA. Dwarf poinciana. Leggy, straggly small tree to ten feet. Grown mainly for its brilliant orange and red flowers. Should not be used as part of shrubbery masses but as a separate small tree for its bloom. Seeds.

POLYSCIAS GUILFOYLEI. Aralia. A common, easy to propagate hedge plant growing to ten feet but with slender form. Should be planted rather closely together for best effects. Several forms propagated in south Florida, including one with white leaf margins and several with finely cut leaves resembling bonsai plants. Relatively free from insects and rapid in growth. Large cuttings may be planted in permanent hedge location. Set back eighteen inches from lot line.

PSIDIUM CATTLEIANUM. Cattley guava. Strawberry guava. One of the finest ornamental large shrubs for Florida landscape design, making a first-rate background plant or screen. Heavy growth of glossy green leaves. White flowers are ornamental, as well as red-purple fruits. Takes pruning well. Its color—a bit of gray in its green leaves—makes it ideal as a background mass for lower plants. Its thriftiness and adaptability make it deal for the home where plants cannot be given constant care. Airlayering or seeds.

PUNICA GRANATUM. Pomegranate. In south Florida this large shrub can be grown only as an ornamental, since it seldom produces fruit of good flavor. Height to eight feet, with six-foot spread. Grown mainly for brilliant orange-red flowers and decorative fruits. Must be grown in good soil and given good care if one expects to maintain an outstanding ornamental. Airlayering or cuttings.

RHAPIS EXCELSA. See Palms.

RUSSELIA EQUISETIFORMIS. Coral plant. Although hardly a true shrub, this vigorous plant will grow to four feet if given suitable conditions. Grows under the most adverse soil conditions, and should be more widely used as a ground cover. Grown mainly for its prolific production of bright red or coral flowers. Leaves are scale-like in size and may not be discerned by a non-botanist. Grow only in full sun for best effects. Cuttings or division.

STENOLOBIUM STANS. Yellow elder. Yellow tecoma. A never failing shrub or small tree to fifteen feet, with spread depending on how skillfully pruned. A great profusion of yellow flowers in winter, even after a severe pruning. Volunteers in vacant lots and pinewoods. Straggly and leggy if not pruned. Seeds or cuttings.

TETRAPANAX PAPYRIFERUS. Chinese rice-paper plant. Grown mainly for the bizarre character of its leaves which are a foot wide. Young leaves are covered with white felt, but as leaves mature only lower surface retains this character. It is a rapid grower and has greener color in part shade than when grown in sun. Produces large clusters of creamy, wooly flowers. Propagated from suckers or from seeds.

THEVETIA PERUVIANA. Yellow oleander. Lucky nut. Makes a small spreading tree if not pruned. Dense foliage makes it excellent for screening or background planting. All parts of this plant, including its fruits, are highly poisonous. Trumpet-like flowers, three inches long, are yellow. Seeds.

THRYALLIS GLAUCA. Golden thryallis. A weak, sprawling shrub to eight feet, irregular when unpruned. Produces a profusion of yellow flowers. Because of its form and the yellow-green color of its leaves this plant may be difficult to grow mixed with other plants. Seeds.

THUNBERGIA ERECTA. Slender shrub which lends itself to pruning and may be trained as a low bushy hedge. Grown mainly for its blue flowers which have yellow throats, produced most of the year. Grows in full sun or in part shade, and thrives on the north side. Cuttings.

TRIPHASIA TRIFOLIA. Limeberry. An armed citrus relative that is widely used for screening and hedges. Grows to ten feet, with almost equal spread. Difficult for intruders to penetrate because of sharp thorns. Excellent background shrub. Withstands pruning well and may be kept low. Fruit dark maroon when ripe. Seeds or airlayering.

VITEX NEGUNDO. Chaste tree. A spreading gray-foliage shrub to twelve feet or more. Grows in sun or semi-shade, but its blue flowers are produced prolificly only in sun. Since vitex makes a large shrub, it should be planted six feet from buildings or walks unless drastic pruning is planned. Sometimes used as a trimmed hedge. Underside of leaf is light gray. Seeds, cuttings, or airlayering.

No other hobby is more rewarding than horticulture. The collecting and growing of rare or interesting plants has attracted physicians and scientists for centuries.

Flowering Tabebuia pallida gives a spring look to Florida

Chorisia speciosa bears its colorful flowers on bare branches

Chapter 6

TREES AND THEIR USES

No garden can be complete without trees. Although shrubs may be the most important plants in any landscape plan, from a practical standpoint, it is hard to imagine an outstanding garden without trees. They provide more than shade. Trees add a feeling of the forest that man seems to need. The most luxurious home, decorated with costly furniture and with all the trappings associated with the best taste and modern affluence, never seems quite right unless there are a few trees about—trees to provide background, to cast their changing shadows over the lawn, and to shed a leaf now and then to remind us that we are not completely apart from nature.

With the advent of the air conditioned home, perhaps shade trees no longer seem so important as they used to be. And it is now possible to insulate a house so well that the planting of trees to protect the roof and walls from the summer sun may no longer be necessary. Yet trees are just as necessary in the landscape as they were before air conditioning, or before it was possible to build the perfectly insulated house. These developments merely give us more freedom in the placement of trees, so that they may serve us better than ever.

Without trees a house stands bare and raw on its bit of land. Our eyes are carried to power poles, to the roof tops of neighboring houses, or perhaps to box-like commercial buildings blocks away. There is nothing to hide an ugly or uninteresting landscape. Trees can be used to frame a house and to set it apart from other houses up and down the street. And trees are especially desirable for creating individuality and charm in a monotonous new subdivision with block after block of the same kind of uninteresting houses.

Space For Trees Limited

Since a tree is meant to be a permanent thing, its selection and planting should be done with great care. On the average city lot there is room for only a few trees. We should avoid planting those which we know will eventually make huge specimens, covering the small lot with solid shade and engulfing the house itself with expanding branches and dense foliage. Such are the large rubber trees *(Ficus benjamina* and *Ficus altissima),* woman's tongue tree *(Albizzia lebbeck),* and the bischofia.

A tree's tolerance to cold and to strong winds should be considered. The spathodea, clusia, and cordia are recommended only for the warmer areas of south Florida. Many trees, especially the flowering ones, shed their leaves in winter and may remain bare of foliage for several weeks. There is no tree, however, which does not shed leaves, either a few at a time throughout the year, or all at once, as do the pongam and mahogany.

In addition there is no tree which, if it attains any size at all, does not develop a root system extensive enough to invade nearby septic tank drain fields. The far-reaching roots of a ficus may fill up septic tank drains in adjoining neighbors' lots.

Fruit Trees Also Ornamental

Tropical fruit trees should not be overlooked as ornamentals. The mango, avocado, sapodilla, lychee, and star apple make as fine an appearance as do many of the non-productive trees. The sapodilla is a robust tree and withstands strong wind about as well as any other wind-resistant tree. Even more striking than the sapodilla is the lychee, which, when covered with bright red fruit, is just as eye-catching as any flowering tree.

While there is no such thing as a hurricane-proof tree, the tamarind probably comes closer to being in this category than any other. Additional trees which tolerate fairly high wind with superficial damage are *Terminalia arjuna, Pongamia pinnata, Albizzia lebbeck, Lysiloma sabicu, Mimusops roxburghii,* bucida, and calophyllum. Even the most storm-resistant tree may be uprooted, however, if it has failed to develop a strong, well distributed root system. When setting out a young tree, all twisted or irregular roots should be cut off. If a young tree has been kept too long in a small container, it may develop a gnarled and twisted root system beyond repair. Such a tree should not be planted, unless it is so rare as to be irreplaceable.

When selecting trees which eventually may reach thirty feet or so in height, it is a good idea to buy trees as large as one can afford. These should be field-grown trees, delivered with a ball of roots covered by burlap, and with their tops cut back somewhat in order to help insure survival. Plantings should by made in well prepared holes, and young trees should be temporarily staked or fixed firmly with the help of stay ropes or guy wires. If wire is used, short pieces of old rubber hose should be cut to cover the end of the wire which is bound around the tree, in order to prevent injury to the bark. A well grown eight-foot tree will develop rapidly and soon reach fair size, whereas if one plants only a switch it may take several years to reach even an eight-foot height.

The placing of trees for best effect is always a problem which deserves a good deal of thought, because after a tree has attained a few years of growth it is difficult to transplant. It is a mistake to set a tree close to a property line. The tree is certain to reach maturity in time and spread its branches over onto neighboring property. Many persons, however, do make this mistake. It should always be kept in mind that your neighbor has a right to cut back to the property line any branches which extend beyond his line. If you have a fruit tree such as a mango or avocado whose branches extend over on a neighbor, it is advisable to share the crop with him, since he has a right to any fruit produced on his side anyway.

Trees should not be set close to the house, so that branches may be dashed against the roof during windstorms. Nor should trees be set near screened patios, because the falling leaves will be difficult to remove from the screen top. Trees planted near swimming pools are almost certain to become a nuisance, even if their branches do not extend over the pool. Leaves may blow for several feet as they fall, or may even be tumbled across the lawn and into a pool.

Trees Require Space to Spread

In general, the placement of trees will depend upon what one really wants. If a spacious green lawn is desired, then several shade trees planted over the yard are certain to destroy the effect. In such instances it is best to plant trees in the background, near the borders. A tree that will eventually make a thirty-foot spread should be planted at least fifteen feet from the lot line.

Nor should trees be massed close together, unless one has in mind a jungle effect. If so, the selection of the trees should be done with care, so that a light shade may be produced for the growing of such jungle plants as philodendrons and anthuriums on the ground and orchids on tree branches. The live oak is by far the best nurse tree for the jungle garden. Those who intend to develop such a garden should search for a building lot on which there are several live oaks of sufficient spread to provide a suitable canopy. Orchids, certain ferns, and bromeliads grow well on the trunk and branches of the live oak. But any tree which produces a fairly open shade and has rough bark might do just as well. Two such trees are the common buttonwood and calabash.

Many a tree is planted because a homeowner saw one in bloom and admired it. This, however, should not be the main reason for planting a tree—because it is striking when in bloom. A tree is in bloom only for a few weeks out of the year. While some trees may be beautiful while in bloom, they may be unsightly at other times, and one may have good reason to wonder whether their short period of glory is worth putting up with their objectionable qualities the rest of the year. Such trees are the bombax, cochlospermum, and the erythrina. Although it is a blaze of glory for a month the poinciana is without foliage for several weeks during late winter and spring; its large black seed pods continue to drop for months after they reach maturity, and every little windstorm leaves a mess behind.

If one is particularly interested in making a collection of flowering trees, it is wise to buy budded varieties when available. These are select trees and likely to be much more dependable than seedling trees, although the cost may be somewhat more. But one should not worry about a little extra cost when so much space on expensive property is given to a tree. The space should be filled by the very best tree the owner is able to obtain—at least in the owner's own estimation.

Below is a list and short description of trees most widely planted in south Florida. No attempt has been made to describe them so that the reader may be able to recognize them on sight. Much effort will be necessary on the part of homeowners in searching for and choosing desirable trees. Many may be seen in nurseries. Others may be studied in parks and botanical gardens. One should not be surprised if a nurseryman has never heard of the name of a tree mentioned. There are many hundreds of tropical trees of which only a few are propagated. Nurserymen propagate the trees for which they get the most demand, or trees with which they are themselves will acquainted and which they particularly admire.

A number of native trees are very desirable for landscape planting, but they are unavailable, or are rarely seen in nurseries. Among the outstanding natives which are seldom propagated by nurserymen, and which we have not included on our list below, are the red stopper *(Eugenia confusa)*, wild cinnamon *Canella (winterana)*, ironwood *(Krugiodendron ferreum)*, and lignum-vitae *(Guaiacum officinale)*. One native tree is included which nurserymen are beginning to propagate because it is easy to grow, and because of its outstanding qualities the pigeon plum *(Coccoloba diversifolia).* It is a common tree of the south Florida hammocks and the West Indies. It is easily propagated by airlayering, is a reasonably rapid grower, and is free of serious pests. Moreover, it is one of the most handsome of our tropical broadleaf evergreen trees and deserves to be carried by landscape nurserymen.

In general, the list below can be expected to serve as only a rough guide to the trees which thrive in south Florida, and especially which one might have reasonable hope of obtaining. Suggested heights that trees may attain refer to trees grown in south Florida. In the more fertile tropics the same trees may attain twice the size they may reach in Florida.

ACACIA AURICULAEFORMIS. Earleaf acacia. A rapid growing willowy tree to about twenty feet. Not wind resistant. Propagated from seeds.

ACHRAS ZAPOTA. Sapodilla. Tough-wooded evergreen tree to forty feet. Widely grown in tropics for apple-size brown fruit. Although the tree is never bare of its leathery leaves, its periodic shedding of a few leaves at a time makes the sapodilla unpopular with those who like a well-swept lawn. Usually grown from seed, but budded varieties are recommended where superior fruit production is desired.

ALBIZZIA LEBBECK. Woman's tongue tree. A very large, spreading tree to sixty feet, with extensive root system, noted for its resistance to strong wind. The tree sheds its mimosa-like foliage in late winter and remains bare of leaves for several weeks. Dry bean-like pods persist, however, and their clattering in the wind has given rise to the tree's common name. Seeds or airlayering.

BAUHINIA BLAKEANA. Blake's bauhinia. Hong Kong orchid tree. A medium-large spreading tree recently introduced from the Orient which has proved to be widely popular because of its long production in fall and winter of large rose-purple flowers which resemble cattleya orchids. Careful training of this tree in its early stage is necessary for the development of a shapely specimen. Sheds foliage in spring. Must be propagated by airlayering, as it does not produce seeds.

BAUHINIA PURPUREA. Fall flowering orchid tree. Medium-large, producing large quantities of purple flowers in the fall. Less desirable than Hong Kong orchid tree. Airlayering or seeds.

BAUHINIA VARIEGATA. Poor man's orchid tree. Medium size, producing large quantities of orchid-like bloom in the spring on bare branches. One of the most striking flowering trees in Florida. Widely popular for parkway planting. A white variety also is available. Airlayering or seeds.

BISCHOFIA JAVANICA. Large evergreen tree which is often mistaken by superficial observers for a Ficus. Rapid growing and withstands periodic flooding. Suitable only where a very large shade tree is desired. Flowers not showy. Airlayering or seeds.

BOMBAX ELLIPTICUM. Pachira. Shaving brush tree. Small tree with brittle branches, grown mainly for its large pink or white flowers which resemble a shaving brush. The flowers, which appear on bare branches, open nightly and persist until mid-day. The coppery red leaves which appear immediately after the flowering period are almost as striking as the flowers themselves. Recommended for parks, botanical gardens, and large estates. Must be grown from cuttings, as seeds are not produced in Florida.

BOMBAX MALABARICUM. Red bombax. Large tree to fifty feet, producing an abundance of red flowers five inches across, on bare branches in winter. Trunk and branches armed with sharp spines. Cuttings.

BUCIDA BUCERAS. Black olive. Upright, well formed tree to forty feet with foliage and conformation resembling a live oak. Wind resistant. Flowers not showy. Widely popular shade tree. Sheds leaves for a short period in spring. Airlayering or seeds.

CALLISTEMON VIMINALIS. Weeping bottlebrush. This is one of the most successful of the several species of the bottlebrush trees planted in south Florida. A somewhat slender but willowy small tree, producing several crops of brilliant red flowers resembling a bottle brush in form. Very striking and ornamental, but more suitable for show and for screening than for use as a shade tree. Airlayering or seeds.

CALOPHYLLUM ANTILLANUM. A very striking evergreen tree with large bright green foliage, reaching thirty feet, with an almost equal spread. Flowers not showy. Produces an abundance of inedible yellow fruits nearly one inch in diameter, thus giving it a reputation for being messy on account of the large amount of fruit dropped. Withstands salt spray. Airlayering or seeds.

CALOPHYLLUM INOPHYLLUM. To casual observer this tree may not be differentiated from *Calophyllum antillanum*. Laurel-like leaves are similar on both trees, but those of *Calophyllum inophyllum* are somewhat larger and thicker than those of *Calophyllum antillanum*.

CANANGA ODORATA. The ylang-ylang. Best grown where a large screening shrub is desired. Although the ylang-ylang does make a fairly large tree in time, it is very brittle and is best held down by pruning to ten or twelve feet in height. Texture of medium green foliage interesting. Often grown for very fragrant flowers, which are not showy but are unusual in form. Seeds.

CASSIA. Shower trees. Several species of the cassia or shower trees are planted in Florida, including *Cassia beareana, Cassia fistula, Cassia javanica, Cassia nodosa,* and *Cassia siamea.* Only an expert botanist or horticulturist can differentiate between some of these. Most popular is *Cassia fistula,* or the golden shower tree, which reaches about twenty-five feet and covers itself with yellow bloom in late spring and early summer. *Cassia grandis,* or pink shower, is a medium-size spreading tree producing large quantities of rose pink flowers in spring. *Cassia nodosa* resembles *Cassia grandis* and is known as the pink and white shower, but when one buys either of these trees one might find later that the tree purchased is a *Cassia javanica. Cassia fistula* is easy to differentiate from the others by its heavy clusters of bright yellow flowers. Airlayering or seeds.

CHORISIA SPECIOSA. Floss silk tree. A relative of the bombax, reaching forty feet in height and producing large numbers of showy rose-pink flowers in spring. The Hatcher chorisia is an outstanding variety of this tree. Airlayering. Hatcher variety is usually grafted onto common chorisia stock.

CHRYSOPHYLLUM CAINITO. See Tropical Fruits.

CHRYSOPHYLLUM OLIVAEFORME. Satin leaf tree. Upright native tree to thirty feet. Leaves copper colored underneath. Tree wind resistant, but despite its attractiveness it is not recommended for planting where its fruit might drop on patios. Growth in early stage is very slow. Seeds.

CLUSIA ROSEA. A spreading tree reaching to about thirty feet, but successfully grown only in warmer places near the coast. It is in the *Garcinia* family, the same as the calophyllum, but the leaves are stiffer and thicker and somewhat less green than the foliage of the calophyllum. Sometimes grown as a large potted plant where unusual character is demanded, such as in a large patio. Tolerates shade. Large fleshy white flowers with pink tint, appearing to be more artificial than real. Seeds or airlayering.

COCCOLOBA DIVERSIFOLIA. Pigeon plum. Striking native hammock tree deserving wider planting for shade or as background.

Rapid growing and easily propagated by airlayering.

COCCOLOBA UVIFERA. Seagrape. Widely popular for its resistance to salt spray, but is also widely grown inland. Its large round leaves are as much as five inches in diameter, and are bright red when new. The common name is derived from its clusters of fruit resembling bunches of grapes. Airlayering or seeds.

COCHLOSPERMUM VITIFOLIUM. A slender tree grown for its bright yellow flowers which are produced on the tips of bare branches in winter and early spring. Single and yellow flowering varieties available. Cuttings.

COLVILLEA RACEMOSA. Upright, to thirty feet, with foliage resembling that of poinciana. Large plumes of scarlet and orange flowers produced in late fall. Because tree does not produce many seeds, it is still fairly rare in Florida.

CONOCARPUS ERECTUS. Buttonwood. Tall or spreading green or gray leaved tree with reclining branches, depending on how tree is trained, or under the conditions it is grown in nature. Excellent tree for growing orchids and bromeliads on its branches. Grown also as shrub. See Shrubs. Airlayering or transplanting from wilds.

CORDIA SEBESTENA. Geiger tree. Small upright tree to fifteen feet, grown for its clusters of yellow or orange flowers. A very tender tree which is best grown near the coast or the Florida Keys. Seeds.

CRESCENTIA CUJETE. Calabash. A small, open tree in south Florida, widely in demand among orchid growers because these plants do so well on its branches. Its fruit, which resembles a gourd, is widely used in the tropics for making drinking cups. Seeds.

DELONIX REGIA. Royal poinciana. Spreading or upright tree to forty feet, and widely planted throughout the tropics as one of the world's most striking trees. The orange or scarlet flowers appear in May and June. The poinciana is looked upon as a messy tree, first dropping its foliage in winter and remaining bare of leaves for many weeks before flowering, and then, after dropping its abundance of flowers, producing large numbers of heavy seed pods which are in turn shed over a long period. It is best grown on large estates, in parks, or parkways. Airlayering or seeds.

ERYTHRINA. Coral tree. Several of the coral trees have been introduced, the best known being *Erythrina variegata.* This is an upright tree reaching thirty feet, producing clusters of scarlet flowers on the tips of branches. Unfortunately it is subject to twig borers which give it a "witch's bloom" appearance. Blooms in spring. Airlayering.

FICUS. Rubber trees. Wild fig. Banyan. Huge, spreading, fast growing trees widely planted where quick effect is desired. Unfortunately these trees become much too large for small homes, and their roots are so aggressive and so extensive that once a large ficus takes over one can expect to have nothing else in a small garden. Several kinds are planted, including *Ficus altissma* and *Ficus elastica*, both of which have large leaves and are incorrectly called banyans. The true banyan, *Ficus benghalensis*, is seldom seen outside of botanical gardens and parks. The most widely planted species are *Ficus benjamina*, also known as *Ficus exotica*, and *Ficus retusa*, usually sold under the name of *Ficus nitida*. *Ficus benjamina* is weeping in habit while *Ficus retusa* is more upright, but both make tremendously large trees. *Ficus religiosa*, or sacred tree of India, is often planted in parks and along parkways. The long pointed, heart-shaped leaves, hanging on long petioles, are kept in constant motion by the slightest breeze, reminding one of the action of aspen leaves in the north. *Ficus lyrata*, or fiddle-leaf fig, sold as a house plant in the north under the name of *Ficus pandurata*, is often planted outdoors in Florida. It eventually makes a medium-large tree, up to forty feet in height. *Ficus sycamorus*, or the sycamore fig mentioned in the Bible, may be seen only in parks and botanical gardens. It produces large clusters of fig-like fruit on the main trunk and on the larger branches. *Ficus carica*, the cultivated fig, is grown in south Florida only in a limited way because its leaves are subject to rust which makes them unsightly. All ficus are easily propagated by airlayering.

GLIRICIDIA SEPIUM. Nurse tree. Madre de cacao. Mother of cocoa. A small spreading but slightly ungainly tree widely planted in the tropics as a living fence or for shading coffee or cocoa. Reaches twenty feet in height and covers itself with pink flowers in spring. Cuttings.

GUAVA. See Tropical Fruits.

HARPULLIA ARBOREA. A rapid growing evergreen tree to thirty feet, popular both for its shade and production of red capsule fruit, which bursts open at maturity, revealing a single black shiny seed inside. Seeds and possibly airlayering.

HIBISCUS ROSA-SINENSIS. Chinese hibiscus. Although the Chinese or red hibiscus, so widely grown for hedges and screens, is usually listed as a shrub, it also can be grown as a small tree.

If so trained it will reach fifteen to twenty feet in height. Cuttings or airlayering.

HIBISCUS TILIACEUS. This gnarled tree grows in thickets in moist areas bordering salt marshes, but it also may be grown as a small tree where quick results are wanted. Its long branches can be trained to take any shape. Cuttings.

JACARANDA ACUTIFOLIA. This tree is hardy to central Florida, where it is a more dependable flowering tree than it is in south Florida. It does best in south Florida following cold winters. Its blue flowers are produced in spring. The mimosa-like foliage produces a light shade. Airlayering or seeds.

KIGELIA PINNATA. Sausage tree. Spreading tree to thirty feet, grown mainly for curious sausage-shaped fruits which reach two feet or more in length. Flowers must be hand-pollinated early in day to insure fruit setting. In native land, Abyssinia, pollinating is done by bats at night. Flowers appear to be sterile to own pollen, so pollen must be brought from a separate tree, as early after daylight as practical because pollen does not retain its vigor very long. Seeds.

LAGERSTROEMIA SPECIOSA. Queen's crape myrtle. Spreads to thirty-five feet, producing dense clusters of pink flowers in early summer. Flowers may be tinted with purple. This tough, wind resistant tree should be more widely planted, but only select trees should be propagated from cuttings or airlayering.

LYCHEE. See *Litchi chinensis* under Tropical Fruits.

LYSILOMA BAHAMENSIS. Common fast growing tree of south Florida and the Bahamas with mimosa-like foliage. Propagated by airlayering, and recommended where a quick effect is desired. Loses leaves for a period during winter.

LYSILOMA SABICU. *Lysiloma latisiliqua.* Horseflesh tree. Cuban lysiloma. A tough, spreading tree to thirty feet, with mimosa-like foliage and producing a light shade. Fast growing and easily propagated by airlayering.

MELALEUCA LEUCADENDRON. Cajeput. A slender tree reaching forty feet in height, with small, narrow leaves and producing an abundance of white flowers three or four times a year. Grows in any soil. Can be made to take a bushy form if training is started in time. Often used as a wind break or background tree, and occasionally as a large specimen "shrub" rather than as a shade tree. When planted in moist places the tree eventually becomes very large and spreading. Seeds.

MANGO. See Tropical Fruits.

MIMUSOPS ROXBURGHII. *Lucuma roxburghii.* An evergreen tree with large stiff leaves, and reaching about twenty feet high. Particularly suitable for screening and for planting near the sea coast, where it is almost as tolerant to salt spray as noronhia. Because it produces an abundance of orange-yellow fruits, over an inch in diameter, it is usually looked upon as a messy tree. Seeds.

NORONHIA EMARGINATIA. A medium-size evergreen tree popular among landscape architects for planting near sea shore. Large stiff leaves resemble those of calophyllum, but are lighter in color. Flowers not particularly showy but very fragrant. Seeds.

MUNTINGIA CALABURA. See Tropical Fruits.

OLEA EUROPAEA. Olive. Although this striking tree does not produce fruit in Florida, it does make a fine small spreading tree. It is very easy to propagate from cuttings or airlayering. Foliage is silvery green. Old trees develop a very unusual character.

PARKINSONIA ACULEATA. Jerusalem thorn. Sometimes called the transparent tree, the parkinsonia is widely used by landscape architects because when planted in front of buildings it does not hide the lines of the building in the background. It reaches fifteen to twenty feet in height with almost equal spread. Yellow bloom add to its attractiveness. Seeds.

PELTOPHORUM INERME. Yellow poinciana. Upright in growth habit, reaching forty feet. The feathery foliage resembles that of the poinciana. Golden yellow flowers are produced in the top of the tree in summer.

PLUMERIA RUBRA. Frangipani. Several species and varieties of frangipani are cultivated in Florida for their clusters of colorful flowers. Tree may reach twenty feet in height but are usually seen as much smaller specimens. Flowers may be white, red, pink or grayish yellow. Large cuttings stuck into the ground will root.

PODOCARPUS MACROPHYLLA. An evergreen conifer reaching twenty feet in height. Used as a sheared hedge. Seeds, softwood cuttings, or airlayering.

POṄGAMIA PINNATA. Pongam. A wind resistant, densely foliaged tree to thirty-five feet, with almost equal spread. Widely used for shade. Sheds leaves in spring, following immediately with new foliage. Flowers not showy, but new bright green leaves in April remind one strongly of spring in the north. Seeds or airlayering.

QUERCUS VIRGINIANA. Live oak. This splendid tree which grows everywhere throughout the southeastern states, is much at home in south Florida. Many so called scrub live oaks are removed from lots cleared for home building which, if left and carefully tended, would make majestic trees. Because of its popularity as a nurse tree for orchids and bromeliads, the live oak may be found in many plant nurseries. Although slow growing at first, the live oak grows fast after about the third year, if pushed with adequate moisture and fertilizer. Seeds.

SCHEFFLERA ACTINOPHYLLA. Queensland umbrella tree. Propagated for many years under the name of *Brassaia actinophylla*, this plant was widely sold for indoor culture throughout the north. It still is, but other plants are now making inroads on its popularity. In south Florida it makes a spreading tree over thirty feet tall. Although used mainly as a shade tree and for its striking evergreen foliage, it does produces brilliant clusters of bright red flowers in the early summer. Airlayering, cuttings or seeds.

SCHINUS TEREBINTHIFOLIUS. Brazilian pepper tree. Christmas berry. Florida holly. The last common name is unfortunate, because this tree is not a holly. It is a fairly small tree in south Florida reaching about twenty-five feet in height with almost equal spread. Bright red fruits are produced in abundance at Christmas time. Unfortunately, some trees produce only male flowers and therefore do not fruit. Usually grown from seeds, but to insure producing a good fruiting specimen it is best to propagate select trees by airlayering.

SPATHODEA CAMPANULATA. African tulip tree. Fast growing tree to about thirty feet, grown mainly for its abundant clusters of large bright red flowers which are produced mainly in summer. Because of its susceptibility to cold damage, this tree should be planted only in the more protected places near the coasts, or in the Florida Keys. It is also subject to damage from high wind. Seeds.

SWIETENIA MAHOGANI. A native tree growing to forty feet in height and widely planted as a street tree. The mahogany loses its leaves for awhile in the spring and immediately puts on a new crop of bright green leaves, which may be eaten by a web-making caterpillar. Prompt spraying with an insecticide will control this pest. Seeds.

SYZYGIUM CUMINI. Jambolan plum. A rapid growing member of the myrtle family, reaching forty feet in height with a thirty-foot spread. Although a striking tree with large evergreen foliage, its abundant production of purple fruit, over an inch in length, makes it impractical as a dooryard shade tree. Fruit edible but has a flavor objectionable to many persons. Seeds.

TABEBUIA ARGENTIA. Silver trumpet tree. Tree of gold. A slender, irregularly shaped tree with light green or silvery foliage, and grown mainly for its golden yellow bloom in spring. Seeds.

TABEBUIA PALLIDA. Pink trumpet tree. A small tree reaching twenty feet in height, and losing its leaves for awhile in spring just before it puts on an abundance of pink trumpet-like flowers on bare branches. Seeds.

TABEBUIA PENTAPHYLLA. Pink trumpet tree. To most persons this tree resembles *Tabebuia pallida,* but it makes a larger tree in Florida, thirty feet or more in height. Seeds.

TAMARINDUS INDICA. Tamarind. A forty-foot tree with almost equal spread at maturity, and noted for its resistance to high wind. Its feathery foliage produces light shade. The abundant pea-like flowers, produced in late spring, are followed by brown pods containing a dark brown, almost candy-like material, in which the seeds are imbedded. The acid-sweet substance is enjoyed by children and is widely used in cooking in the tropics. Seeds.

TERMINALIA CATAPPA. Tropical almond. Large tree reaching forty feet, noted for the autumn colors of its large foliage after the onset of cool weather in south Florida. The leaves turn a bright red before dropping. Seeds.

TERMINALIA ARJUNA. Wild almond. This spreading tree, reaching forty feet, is so different in appearance from its relative, *Terminalia catappa,* that only a botanist would note the similarity. It is widely planted as street tree because of its form, its resistance to pests, and also because of its resistance to high wind. Seeds.

Rampant Congea keeps its pink color for weeks in spring

Individual flowers of Solandra longiflora are spectacular

Chapter 7

FLOWERING CLIMBERS

Tropical vines give Florida much of its color. There is no time of the year when one or more climbers are not in bloom. All put together – the bougainvillea, allamanda, congea, and solandra – you might think they would make a gaudy collection of colors indeed. Yet, scarlets, pinks, yellows, purples, and blues are planted throughout south Florida gardens, and, somehow, in their masses of bloom, only the sensitive, color-conscious person notices the clash. However, a more careful selection of color combinations would improve these gaudy gardens.

The most widely grown of the vines in Florida is the bougainvillea. It probably owes much of its popularity to the fact that it puts on its showest bloom in winter. Next in popularity is the yellow allamanda, covered with its large bloom from spring until fall. Also popular is the flame vine, running over the roofs of many Florida homes, and putting on its flame colored flowers in early spring.

There are vines for every purpose and vines with flowers of every color. Vines can be used in many ways, on walls and fences, on pergolas and porches where shade is wanted. Many vines, like the beaumontia, cryptostegia, and the Rangoon creeper (*Quisqualis indica*) have luxurious foliage and are suggestive of the lush tropics when grown in areas where they have freedom to develop. Most vines, however, do not thrive in shady areas, or if they do, they bloom very poorly. In such areas, as in jungle gardens, the flowering climbers should be avoided, and such climbers as the philodendron and pothos should be planted.

A large number of vines can be trained as shrubs, including the several jasminums, cape honeysuckle, cryptostegia, Chinese hat plant, and even the allamanda. A number of vines may be used as ground covers, such as the jasminums again, Confederate jasmine (*Trachelospermum jasminoides*), as well as a shrub type of allamanda, *A. nerifolia.* Most vines require something to run on, such as a fence or a trellis, and they do their best flowering when they are off the ground.

Trellises are often made of wood, but wood decays rapidly in south Florida and should not be used unless it has been pressure-treated. Even so, it can be expected to last only ten to fifteen years when in contact with the soil. Where vines are to run over arbors, it is best to have the upright supports made of poured concrete rather than creosoted posts. The concrete posts can be painted and be further covered by having vines trained up on them. Horizontal supports for the vines can be made of pressure-treated poles. It must be remembered that vines become very heavy in time and that very strong support is needed.

Woody vines, such as the bougainvillea and petrea, can be trained on walls by using aluminum wire to attach them to concrete nails previously driven into the concrete. Vines should not be permitted to cover wooden walls in south Florida, where long periods of dampness are likely to occur during the summer months.

Vines do require pruning, but pruning in itself does not insure bloom. Probably the best time to do any heavy pruning is just after a vine is through flowering. Vines which bloom in winter and spring should certainly not be pruned heavily late in the year. August should be about the last month for any pruning of these vines, among which is the bougainvillea. Probably the best way to insure the flowering of bougainvillea is to withhold both fertilizer and irrigation during the fall and winter, then irrigate when it is evident that the vines will have a full crop of flowers. Pruning can be done in late spring or early summer, followed by one of two applications of fertilizer to promote new growth. The form of the new growth can be controlled by careful pruning.

The cultural requirement of vines is no different from those of shrubs and trees. And many vines, like the shrubs and trees, are subject to mineral deficiencies. One or more applications a year of chelated iron to the allamanda and congea will help to keep their leaves in healthy green color.

Below is a list of the more common vines grown in south Florida. None listed here is really rare in Florida, but few vines are propagated by nurserymen. Thus it may be difficult to find more than a few of them in plant nurseries. Easy to obtain should be the allamanda, bougainvillea, jasminum, petrea, pyrostegia (flame vine), and the thunbergia.

ALLAMANDA CATHARTICA. Yellow allamanda. A sprawling vine which may be trained on trellises or fences. It has large glossy leaves and produces golden yellow, bell-shaped flowers up to four inches in diameter. Variety Hendersoni is the one most widely planted. Flowering of well grown plants lasts from spring until fall, with a scattering of bloom even in winter. A second variety, Williamsi, produces much smaller flowers and blooms more sparingly than Hendersoni. Airlayering or cuttings.

ALLAMANDA NERIIFOLIA. More of a shrub than *Allamanda cathartica.* This species tends to bloom in late winter and spring. Propagated from cuttings and from seeds when produced.

ALLAMANDA VIOLACEA. Purple allamanda. A sprawling, shrubby plant which can be trained as a vine. Often confused with the rubber vine, cryptostegia. Layering or cuttings.

ANTIGONON LEPTOPUS. Coral vine. An aggressive climber producing large quantities of rose-pink flowers periodically throughout the warm season. Vine becomes unsightly after flowering ends in late fall, and may be cut back. Seeds freely and tends to become a weed.

ARGYREIA SPECIOSA. Wooly morning glory. A large climber, with heart-shaped leaves up to ten inches in diameter, and which are covered with white felt beneath. The rose-purple morning glory type flower has a deep purple throat. Blooms in spring and summer. Layering or hardwood cuttings.

ARISTOLOCHIA. Several species of this vine with its unusual flowers are grown as oddities in Florida. One species is known as pelican flower because its flower resembles the bird for which gets its name. Another species is known as rooster flower. Some of the aristolochia flowers attract flies by giving off the odor of carrion. Cuttings or seeds.

BAUHINIA GALPINI. A sprawling shrub which can be trained as a vine. See **Shrubs.**

BEAUMONTIA GRANDIFLORA. Trumpet vine. An immense vine growing to great heights, eventually producing a woody trunk several inches in diameter. This rampant climber is a handsome vine at all times of the year, and is quite suited for pergolas where a heavy shade is desired. Its large white flowers, the size and shape of Easter lilies, are produced in spring. Airlayering.

BIGNONIA MAGNIFICA. (*Arrabidaea magnifica.*) A sprawling shrub or vine with dark green, leathery leaves, and producing rose-purple flowers, a few at a time throughout the warm months. Cuttings.

BOUGAINVILLEA GLABRA. Two varieties, Sanderiana, purple in color, and Crimson Lake, crimson in color. Sanderiana may be grown as a sprawling shrub or sheared shrub. Crimson Lake is more of an ungainly plant and does not take shearing. Its flowers are produced along vigorous, heavy shoots which, if cut back late in the fall, prevents the plant from coming into satisfactory bloom. The variety Afterglow is similar to Crimson Lake in form but produces yellow-orange flowers. The varieties of *Bougainvillea glabra* may bloom periodically throughout the year, but their heaviest bloom is likely to be in late winter or spring. Cuttings or airlayering.

BOUGAINVILLEA SPECTABILIS. An immense vine, capable of reaching the top of a four-story building if trained on the wall. Variety Lateritia, or Brick Red, is bright red in color. Blooming, however, usually takes place but once a year, in the winter, when the vine is completely hidden by its colorful flower bracts. Although there are several varieties of bougainvillea in the nursery trade, it has been difficult to place them in their proper species. Buyers should consult nurserymen about a plant's growing and flowering habits before purchasing it. Hardwood cuttings or airlayering.

CLERODENDRUM THOMSONIAE. Bleeding heart. This well known vine produces large clusters of white bag-shaped flowers which appear to bleed scarlet at their tips. Actually the individual flower is scarlet and is almost covered by a bagshaped calyx. Blooms in spring and summer. Cuttings.

COMBRETUM GRANDIFLORIUM. Showy combretum. A vigorous grower, producing bright red flowers almost brush-like in form. The new leaves, copper in color, are almost as striking as the flowers. Blooms in spring. Cuttings or layering.

CONGEA TOMENTOSA. A heavy, strong climber covering itself in late winter and spring with pink bloom of changing tint. Actually it is the bracts which stay on the vine for several weeks after the individual flowers have finished blooming. Because the congea is affected by iron deficiency when grown in limestone soil, it is helped by one or more applications a year of a chelated iron. Layering.

CRYPTOSTEGIA GRANDIFLORA. Rubber vine. A vigorous, sprawling vine which may be trained as a shrub. It has funnel-shaped purple flowers which are produced in small numbers throughout the year. This vine is grown mainly for its large glossy leaves. It appears to have few insect enemies. Layering or cuttings.

CYDISTA AEQUINOCTIALIS. Garlic vine. An attractive evergreen vine producing panicles of pink flowers in spring and in summer. The crushed leaves and flowers have the odor of garlic. Seeds.

DERRIS SCANDENS. Malay jewel vine. This tremendous climber is capable of growing to the tops of the tallest jungle trees, producing a trunk itself a foot or more in diameter. It is covered with clusters of fragrant white flowers periodically during the warm season. Hardly suitable for the small garden. Seeds.

DOXANTHA UNGUIS-CATI. Cat's claw vine. Climbing with claw-like tendrils, this vine can scale any wall. The vine produces an abundance of bright yellow, trumpet-shaped flowers in the spring, but they last only a few days. The seeds become widely distributed and this vine is likely to become a weed.

FICUS PUMILA. Creeping fig. An evergreen vine, growing on walls and producing an ivy-like effect. Not recommended for planting about the home.

HOLMSKIOLDIA SANGUINEA. Chinese hat plant. A sprawling shrub which may be trained as a vine. Grown for its numerous brilliant orange-red flowers that resemble old style Chinese or Mandarin hats. Flowers much of the year, including winter. Cuttings.

HOYA CARNOSA. A slow growing vine which can be grown only in soil which is free from nematodes. Grown for its clusters of waxy, star-shaped flowers. It is best grown on the trunk of a cabbage palm, where it can establish itself with the aid of aerial roots which invade decaying leaf bases. An occasional application of soluble fertilizer to the plant and to the palm trunk promotes growth. Cuttings.

IPOMOEA. Morning glory. Although there are several species of morning glories, familiar to most gardeners, one of the finest of the group, *Ipomoea horsfalliae,* is seldom planted. It is a vigorous grower and is one of the few vines which blooms well on the north side of a building. The numerous clusters of purplish-red flowers are produced several times a year, even in winter. This species does not seem to produce seeds in south Florida, and must be propagated by airlayering. The wood rose, *Ipomoea tuberosa,* is more widely planted as an oddity than as an ornamental. It is an aggressive vine, producing bell-shaped yellow flowers. The brown seed pod, which develops later, is surrounded by open sepals which give the appearance of a dried rose. Seeds.

JASMINUM. Several species, but grown mainly as sprawling shrubs. See **Shrubs.**

PASSIFLORA. Several species of this twining vine, grown for their edible fruits as well as for their unusual flowers. *Passiflora coccinea* produces a brilliant red flower. Seeds or layering. See **Tropical Fruits.**

PETREA VOLUBILIS. Purple wreath. A very popular vine in southern Florida, reminding some persons of wisteria which does not thrive in south Florida. Especially suitable for a trellis. The purplish blue flowers are borne in racemes and remain on the vines for several weeks. Flowering season is spring and summer but sometimes a vine may produce flowers at any time of the year. Airlayering.

PODRANEA RICASOLIANA. A climber with open, small foliage and producing clusters of bell-shaped pink flowers striped with red. A winter and spring bloomer. Cuttings or layering.

PORANA PANICULATA. Christmas vine. A tall climber, bearing immense sprays of white flowers in November and December. Seeds or layering.

PYROSTEGIA IGNEA. Flame vine. A widely grown twining climber bearing a profusion of bright yellow tubular flowers in spring and early summer. The orange-yellow flowers persist for several weeks. It is often seen growing on the roofs of homes in southern Florida. Layering.

QUISQUALIS INDICA. Rangoon creeper. An aggressive shrubby vine suitable for growing on arbors or stone walls. It will pile up on a stone wall and form a dense, shrubby growth. Clusters of pink to red flowers are produced in the summer. Layering.

SENECIO CONFUSUS. Mexican flame vine. This scraggly vine produces masses of orange-red flowers mainly in spring and summer. Individual flowers are daisy-shaped and are about one inch in diameter. Seeds or layering.

SOLANDRA LONGIFLORA. Bugle vine. Chalice flower. A heavy, woody climber with glossy, green leaves and producing large trumpet shaped flowers nearly a foot in length. Summer and fall. Airlayering.

SOLANUM SEAFORTHIANUM. Brazilian night shade. A small, open vine grown for its clusters of blue or purple-blue flowers about an inch across. The flowers are followed by clusters of small fruits, turning from green to red and then to black at maturity. The fruits are enjoyed by birds. Seeds.

SOLANUM WENDLANDI. Costa Rican night shade. This is a vigorous grower, producing large clusters of blue flowers in spring and summer. Individual flowers are up to two inches in diameter. Seeds.

STEPHANOTIS FLORIBUNDA. An evergreen vine grown for its delicate, fragrant, creamy white flowers, which appear throughout the summer. This vine must be grown in nematode-free soil. Air layering.

TECOMARIA CAPENSIS. Cape honeysuckle. A shrubby vine producing clusters of brilliant orange-red flowers, mainly in spring and summer. Often grown as a shrub. Cuttings.

THUNBERGIA GRANDIFLORA. Bengal clock vine. A rapid growing, tall climber especially suitable for arbors. The blue, bell-shaped flowers appear at all months of the year. The white flowering variety also is in cultivation. Layering.

TRACHELOSPERMUM JASMINOIDES. Confederate jasmine. This evergreen vine covers itself with highly fragrant white star-shaped flowers in spring and early summer. Layering.

Palms give tropical lushness to Florida or Bahamas gardens

Cluster palms are widely used in many landscape settings

Chapter 8

PALMS

Florida is rightly called the palm state. Palms dominate the scene throughout much of the peninsula. Large colonies of picturesque sabals stand for miles along the sand ridges of the east coast and occupy large areas of flatwoods in central and south Florida. Saw palmettos make up the understory scrub through miles of pinewoods. Coconut and royal palms line the streets and waterways of south Florida. Clusters of slender native thrinax and silver palms adorn the keys from the mainland to Key West. Florida has eleven palm species.

How many kinds of palms there are has been estimated at various times between 1,000 and 6,000 species. These plants are scattered throughout the tropics, often in almost inaccessible places, where botanists have been unable to collect them. The most recent estimate of palm species was made by Dr. Harold E. Moore, Jr., Cornell University palm authority. In *Principes,* journal of the Palm Society, he estimates the number at 2,640 species.

No other family of plants has more interesting, varied, and graceful members than the palm family. In Florida we usually think of the palm mainly as an ornamental, but it is one of the world's most important food sources, furnishing a large percentage of the population of the tropics with food, oil, and materials for building shelter. Millions of dollars worth of palm products, including oils, coconut meat, and fibers, are exported from the tropics to the other parts of the world.

The largest collection of palm species in the Western Hemisphere is at the Fairchild Tropical Garden, in Coral Gables, Fla., where close to 500 kinds have been planted.

Although many palms grow in warmer areas of the Temperate Zone – *Trachycarpus fortunei* and *Sabal palmetto* are hardy as far north as southern Virginia – the palm is usually associated with the tropics. The palm is cultivated throughout the world where it is obtainable – outdoors in warm climates and indoors in the cold north.

Palms vary greatly in size and form. There is a palm for every purpose in the design of a tropical garden. Is there another tree which could match the royal palm for formal planting, with its tall columnar trunk? And what other tree could match the graceful coconut, with its curving trunk and long, feather-shaped fronds? If it is color one is seeking, there is nothing more striking than the Merrill palm (*Veitchia merrilli*), with its clusters of bright scarlet fruits at Christmas time. And for spectacular, massive production, there are no flowers which can surpass the tremendous bloom clusters of a sugar palm (*Arenga pinnata*) or the fishtail palm (*Caryota urens*).

Palms vary in size from the four-foot *Neanthe bella* (*Chamaedorea elegans*) to the tall royal which may reach sixty to eighty feet. Both shade-loving and sun-loving palms are available for the landscape designers. Cluster palms are available for screening, including *Rhapis excelsa, Paurotis wrightii, Chrysalidocarpus lutescens, Phoenix reclinata, Chamaerops humilis,* and *Nipa fruiticans.* The nipa palm lives in brackish swamps with mangrove and buttonwood. *Gaussia princeps,* which thrives among the rocks of the haystack hills of western Cuba, may start out as an epiphyte, just as a strangler fig may start out on the trunk of a cabbage palm or on the branch of a live oak. The gaussia may start in a crevice between rocks fifteen feet or more above a ledge of soil, where it grows just as an orchid would grow until after several years when its roots make contact with the soil below. The slender *Gaussia princeps* thrives in South Florida and deserves wider planting because of its adaptability to limestone soil, as well as for its colorful clusters of red fruit. The phoenix or date palms like the open where they receive sun most of the day. Most other palms, however, will thrive in part shade.

In choosing palms for planting in a small garden, species should be selected which will not grow out of control. A palm cannot be kept small by pruning. A Canary Island date palm eventually makes a large specimen, and nothing will hold it back except time and poor growing conditions. Likewise, a coconut planted beneath power lines eventually will grow into the wires where its fronds must be cut back, thus reducing this noble palm to a monstrosity. Palms which die after flowering also should be avoided, unless one has a very large place or is making a special collection. *Caryota urens* and *Arenga pinnata* may be striking palms when they bloom. But after progressively flowering from top to bottom, which takes three to four years, the palm dies and must be removed. To get rid of a dead arenga thirty feet high might not be so easy, if this palm is growing among one's ornamentals in a small garden.

Palms may be used in many ways in the landscape. Like shrubs, they may be used for screening or for background. Like trees, the larger palms may be used for shade. As a texture plant the palm is unsurpassed. The landscape designer can find palms to solve almost any problem in the garden. But how to use palms well with other plants, such as young palms with shrubbery, can be a difficult problem. However, shrubbery may serve well as understory plants after palms have developed trunks several feet high.

In the first place, however, it is generally best to plant together palms of similar character. For instance, many kinds of shrubbery or ground covers may be used at the base of a colony of slender palms, such as *Veitchia merrilli, Veitchia winin, Dictyosperma album, Ptychosperma elegans,* and the several species of *Coccothrinax* and *Thrinax.* Where colonies of the native cabbage palm exist on a building lot, every effort should be made to preserve as many as possible. They make an excellent place to build a jungle garden, with many kinds of orchids and bromeliads planted on the trunks, and with philodendrons and anthuriums planted in the soil at their base. Large single trunked palms, such as the Canary Island date palm, and the slender *Ptychosperma elegans,* should never be planted close together. They simply do not make a desirable or attractive association. One or the other will seem out of place.

Just as with trees and shrubbery, palms should not be planted throughout the lawn, like candles on a green cake. Palms are best restricted to borders, except for their charm and shade about patios or where they are to be used for some special purpose, as in the development of a jungle garden.

Palms can be hazardous. A coconut can cause serious injury if it falls on one's head. A coconut can also make a sizeable dent in the body of an automobile, or if it is planted close to a house, the falling nut can break the tile of a roof. It may seem odd to think of anyone being injured by a falling leaf, but the fronds of many palms are very large and heavy. Thus large palms should not be planted where they are likely to shed their large fruits or leaves on people sitting beneath them. One has little cause to worry from palms such as coccothrinax, thrinax, ptychosperma, arecastrum, phoenix, sabal, or washingtonia.

Most palms are grown from seeds. Only one species, rhapis, is generally propagated by division. Palm seeds vary from pea-size to the huge double-coconut, the largest seed in the world, which resembles two coconuts grown together. This palm, a native of the Seychelles, unfortunately does not thrive in Florida. Thus the coconut is our largest seed. When planting the coconut do not remove it from its husk. To sprout coconuts, several of them can be placed on the ground close together and covered with wood shavings, grass clippings, or any other material which will hold moisture. Play a sprinkler on the bed two or three times a week. A few sprouts should begin to show after several weeks. After the nuts have sprouted they may be planted in permanent locations.

Other palm seeds may be planted in clay pots or seed flats, and should be slightly covered with the planting material. An excellent material for planting seeds other than the coconut can be made by mixing equal parts of coarse sand and perlite or crushed granite, to which a small amount of shredded sphagnum is added. No fertilizer need be added until after the seeds have sprouted. The large seeds of the gingerbread palm (*Hyphaene thebaica*) and borassus should be planted in the location where these palms are to remain permanently, as they are very difficult to transplant or to get started in containers. Some palms grow so slowly that it may be necessary to keep them in containers for four to six years before planting them in permanent locations.

Palms, like other plants, require fertilizer and water to grow well in south Florida's poor soil. Palms must also be protected from insects and diseases. Many of them require special treatments to help them overcome nutritional deficiencies. Palms whose fronds turn yellow despite regular applications of fertilizer may be helped by sprays or drenches containing zinc, manganese, and chelated iron. However, palms which require shade will turn yellow when grown in full sun, and no amount of sprays or drenches will make them develop green leaves.

Palms, more than most other plants, must be grown where they can reach full development. They are usually most effective when planted in the background, with sufficient open space left in front of them so that one can enjoy the effects, both from a distance as well as a close up view. There are no more magnificent plants in the world than tthe palms, which deserve the family name, *Principes,* given to this group by the father of modern botany, Linnaeus.

Below is a list of the more common palms and some of the rare palms which are grown in South Florida. Most of them are available, but the rare species may be found with some difficulty:

ACROCOMIA. Several species of heavy, spiney-trunked, feather-leaved palms, twenty to sixty feet, including *Acrocomia aculeata, A. armentalis* (has bulging trunk), *A. mexicana, A. sclerocarpa,* and *A. totai.* The last named is one of the hardiest of palms, withstanding temperatures as low as 14 degrees.

ADONIDIA. See *Veitchia.*

AIPHANES. Several species of slender spiney-trunked, feather-leaved palms. Grown for colorful red fruit clusters. Not hardy.

ARCHONTOPHOENIX ALEXANDRAE. Alexandra palm. Slender, graceful palm with feather leaves, to forty feet. About equal to royal palm in hardiness.

ARECA. See *Chrysalidocarpus lutescens.*

ARECASTRUM ROMANZOFFIANUM. *Cocos plumosa.* Feather-leaved palm widely planted throughout southern half of Florida, reaching thirty-five feet.

ARENGA PINNATA. Sugar palm. Short lived single-trunk palm with feather leaves, reaching thirty feet. Sap from fruit causes severe itching if it comes into contact with skin. Handsome palm, but dies after flowering. Large specimens may be seen at Fairchild Tropical Garden.

ASTROCARYUM. Several species of this sturdy, feather-leaf palm. Slow growing but worth planting. Rare in Florida.

ATTALEA. See *Orbigyna.*

BACTRIS GASIPAES. Peach palm. Formerly known as *Guilielma gasipaes.* A spiney, multiple-trunked palm, yielding plum-size yellow fruits which are used as food in Central America. Thirty feet. Rare in Florida.

BISMARCKIA NOBILIS. Large, single-trunk, fan-leaf palm not yet widely known in Florida. Sometimes grown under name of medemia.

BORASSUS FLABELLIFER. Palmyra palm. One of world's most important and imposing palms. Its single, large trunk reaches nearly one hundred feet, with huge, black, fan-shaped fronds. Rare in Florida.

BUTIA CAPITATA. Hardy, slow growing palm thriving as far north as south Georgia. Fruit edible.

CARYOTA MITIS. Fishtail cluster palm. Multiple-trunk palm reaching thirty feet, widely used in south Florida landscape planting.

CARYOTA URENS. Fishtail palm. Single-trunk palm reaching forty feet, and dying after it finishes flowering.

CHAMAEDOREA. Several species of this palm, best known of which are *Chamaedorea elegans* (*Neanthe bella*), *C. erumpens* and *C. seifrizii.* A number of other chamaedorea species thrive in Florida and deserve wider planting, including *C. geonomiformis, C. graminifolia, C. brachypoda,* and *C. tepejilote.* All chamaedorea species require shade.

CHAMAEROPS HUMILIS. A slow growing cluster palm which thrives in southern California and throughout Florida. Rare in Florida, where it seldom produces seeds.

CHRYSALIDOCARPUS LUTESCENS. Incorrectly called areca palm. Popular cluster palm throughout south Florida, with many yellow-green trunks displaying prominent rings. Yellow-green leaves are feather-like. Two other cluster types of *Chrysalidocarpus* palms are planted in Florida, but are rare. They are *C. madagascariensis,* and also the cabada palm which is now listed only as *Chrysalidocarpus species,* because its origin is as yet unknown. It appears to be well adapted to south Florida soils and climate and promises to be a popular palm. Its multiple trunks reach forty feet.

COCCÓTHRINAX. Several species of this slow growing palm from Florida and the West Indies, some of them being among the most striking of the palm family. All have fan leaves which are silvery underneath. They include *Coccothrinax crinita,* whose trunk is covered with fiber; *C. argentata,* the silver palm of the south Florida pinewoods and the Florida keys; *C. Argentea,* the West Indian silver palm, and *C. miraguama,* a graceful Cuban silver palm.

COCOS NUCIFERA. Coconut. Several varieties, including the common coconut, dwarf green Malaya, dwarf orange Malaya and dwarf yellow Malaya. The Malaya dwarf varieties are said to be resistant to the nematode which has killed most of the common coconut trees in Key West. The dwarf varieties come into production earlier than the common coconut, but eventually grows about as tall as any other coconut.

COPERNICIA. Several species of this interesting palm, mainly from Cuba, are in the Fairchild Tropical Garden collection. They are generally very slow growing and are not likely to be found in any but nurseries specializing in palms. All are single trunked with fan-shaped leaves.

CORYPHA UMBRACULIFERA. Talipot palm. One of the largest of palms, seldom seen in Florida outside of palm collections. Reaches forty to sixty feet, then produces one immense inflorescence at the top and dies.

DICTYOSPERMA ALBUM. Hurricane palm. A graceful slender palm with feather-like leaves, reaching thirty feet, bending easily in high winds.

ELAEIS GUINEENSIS. African oil palm. A large feather-leaf palm, important in tropics for its production of palm oil. Available but seldom planted.

GAUSSIA PRINCEPS. A slender palm from western Cuba apparently well adapted to Florida's limestone soil. Reaches twenty feet and yields clusters of bright red fruit. Rare in Florida.

GUILIELMA GASIPAES. See *Bactris gasipaes.*

HETEROSPATHE ELATA. Tall, slender palm with feather-like leaves. Grown in warmer areas along the coast, but rare.

HOWEIA. Two species of this graceful, slender-trunk palm with feather-like leaves, *Howeia belmoreana* and *H. forsteriana.* Both thrive in shade.

HYPHAENE. Called gingerbread palm because fruit resembles gingerbread in color and flavor. Several species, of which *Hyphaene thebaica* is the best known. A few other species of *Hyphaene* introduced by the Fairchild Tropical Garden are not yet in cultivation.

LATANIA. This single-trunked, fan-leaved palm is represented by three species, *Latania borbonica,* the red latan palm; *L. loddigesii,* the blue latan palm, and *L. verschaffeltii,* the yellow latan palm. These palms reach thirty feet in Florida, but are restricted to the warmer areas along the coasts and in the Keys. Male and female flowers are borne on separate palms.

LICUALA GRANDIS. A small, slender-trunked palm with very graceful, round fan-shaped leaves. Does well in shade, but it is very tender. Rare in Florida.

LICUALA SPINOSA. A cluster palm with divided fan-shaped leaves, thriving in shady locations. Rare in Florida.

LIVISTONA CHINENSIS. Chinese fan palm. A common single-trunked palm with large fan-shaped leaves resembling the foliage of the common sabal. Hardy throughout much of Florida. Bears large clusters of blue-green fruit.

LIVISTONA ROTUNDIFOLIA. Very slender single-trunked palm and fan-shaped leaves. Prominent rings on trunk. Will grow in sun or part shade. Somewhat rare but a promising palm for South Florida.

MASCARENA. Hyophorbe. An odd, thick-trunked palm represented by two species in Florida, *Mascarena legenicaulis,* the bottle palm, and *M. verschaffeltii,* the spindle palm. Not rare, but grown only as an oddity.

NEODYPSIS DECARYI. A most unusual and striking tringular-shaped palm introduced by the Fairchild Tropical Garden but not yet in wide cultivation. A very promising palm for south Florida.

NIPA FRUITICANS. A cluster palm growing in brackish mangrove swamps. Although rare in Florida, it thrives here and deserves wider planting.

ORBIGNYA. A very large, slow growing palm which looks much like the attalea and scheelea. Its feathery leaves may reach thirty feet in length. Rare in Florida.

PAUROTIS WRIGHTII. A common but slow growing cluster palm native to the Everglades National Park and Big Cypress Swamp. Its multiple, slender trunks eventually make a large clump. The leaves are fan-like. Generally transplanted from the wilds. Slow from seeds.

PHOENIX CANARIENSIS. Canary Island date palm. A large palm with massive trunk and spreading top with feather-like leaves. Hardy to south Georgia.

PHOENIX DACTYLIFERA. Date palm. Although this famous palm, cultivated for thousands of years, will grow in Florida it will not produce good fruits here. It requires a dry climate for dependable fruit production. Best varieties propagated from suckers.

PHOENIX RECLINATA. A cluster palm with many slender trunks, widely popular among landscape architects. Like most other date palms it will grow throughout the Florida peninsula.

PHOENIX ROEBELENII. The smallest of date palms, reaching six to eight feet in height, with delicate feathery leaves.

PRITCHARDIA PACIFICA. A slender palm with stiff fan-shaped leaves reaching thirty feet. Very tender to cold. A similar species, *Pritchardia thurstonii,* is also planted in south Florida.

PSEUDOPHOENIX. A distinctive palm with gray-green color in its single trunk and feather-like stiff leaves, represented by three species in Florida, *Pseudophoenix saonae, P. sargentii* and *P. vinifera. P. sargentii* is believed to be a native palm of the Florida Keys, and is found throughout the Bahamas and West Indies. The species of pseudophoenix are very slow and are seldom planted except in collections. These palms are easily transplanted if root-pruned some weeks in advance of moving.

PTYCHOSPERMA ELEGANS. A slender graceful palm with feather-like leaves, reaching thirty feet. Bears large clusters of bright red fruit.

PTYCHOSPERMA MACARTHURI. A cluster palm with many slender trunks, widely used in landscape planting. Bears large clusters of bright red fruit.

PTYCHOSPERMA SPECIES. Taksi palm. A slender, graceful palm from the Solomon Islands, introduced to Florida by Dr. David Fairchild. Much like *Ptychosperma elegans*. Still rare but a promising palm.

RHAPIDOPHYLLUM HYSTRIX. Needle palm. A sprawling cluster palm with bright green fan-like leaves, and armed with long black spines near the leaf bases. Becoming popular in landscape planting, the plants being removed from the wilds and trucked to south Florida.

RHAPIS EXCELSA. Bamboo palm. A slender suckering palm with dark green, deeply cut fan-like leaves, growing to twelve feet or more. Used for screening, and especially desirable where there is part shade. Hardy to north Florida.

ROYSTONEA REGIA. Royal palm. One of the world's most noble palms, widely planted throughout the tropics. Reaches forty to seventy feet. Botanists list several other species of *Roystonea,* including *Roystonea oleracea,* the cabbage royal of South America and Trinidad. This species is said to reach one hundred to one hundred thirty feet.

SABAL PALMETTO. This is Florida's best known native palm. It covers large areas of the east coast and the flatwoods of central and southern Florida. Under favorable conditions it will reach sixty feet or more but in extreme south Florida is seldom seen over thirty feet. Many thousands of these palms are transplanted from the wilds each year and used in landscape planting. A dozen other Sabal species may be seen at the Fairchild Tropical Garden.

SERENOA REPENS. Saw palmetto. Locally known as scrub palmetto. A common low growing palm with fan-like leaves in the Florida pinewoods. Recovers quickly after being periodically burned over. When protected from fire this palm develops an upright trunk. Difficult to transplant. Because this palm is so common it is not looked upon as an ornamental, and is usually destroyed when building lots are cleared. On the other hand, the saw palmetto's ornamental qualities may surpass those of many of the exotics which homeowners use to replace it.

SYAGRUS. A slow growing palm of Central and South America, seldom propagated as an ornamental.

THRINAX. Two species of this slender fan palm are native of Florida, *Thrinax microcarpa* and *Thrinax parviflora. T. microcarpa* is much the slower grower and may be differentiated from the other thrinax by its gray-green leaves. *Thrinax parviflora* is a much taller palm at maturity, and its leaves are a glossy green.

TRACHYCARPUS FORTUNEI. One of the hardiest of all palms, growing throughout the southeastern states and as far north on the West Coast as the Canadian border. Slow growing, with single trunk and fan leaves.

VEITCHIA. Two newly discovered species of *Veitchia* palms from New Hebrides recently have been added to this group of popular ornamentals. *Veitchia merrillii,* formerly *Adonidia merrillii,* has been in cultivation for several years in Florida and is now widely planted for its clusters of bright red fruit, which mature at Christmas. The two new species, recently named, are *Veitchia montgomeryana* and *Veitchia winin.* These are all very fast growing palms and are especially recommended for the small garden.

WASHINGTONIA ROBUSTA. This palm, a native of Mexico, is commonly grown throughout Florida. Another species, *Washingtonia filifera,* a native of California, is almost rare in Florida. The two palms resemble each other while young, but a mature *W. filifera* has a stout trunk up to three feet in diameter near the base. Both palms are distinguished by their habit of retaining their dead leaves, which may persist for several years. The name honors George Washington.

ZOMBIA ANTILLARUM. A slow growing cluster palm with many slender trunks armed with sharp spines. Its fan-shaped, deep green, glossy leaves make this palm popular among collectors, but its growth is so slow that nurserymen seldom carry it.

Plant form and leaf texture are as important details in a garden as is color. To develop and maintain the kind of garden which is nearly always filled with bloom requires a great deal of time and energy, or great cost if one hires the work done. For this reason we usually find more neglect in flower gardens than in green gardens.

Choquette is one of many popular Florida avocado varieties

Lychee is one of tropics' most colorful fruiting trees

Chapter 9

TROPICAL FRUITS

Tropical fruit producing trees and shrubs are among south Florida's outstanding ornamentals. Citrus is, of course, the most widely planted ornamental tree in the State. In the warmer parts of south Florida a large number of tropical and semi-tropical fruit bearing plants can be grown. Most of the tropical fruit trees are broadleaf evergreens, holding their foliage all year. Several of them, especially the eugenias, cattley guava, macadamia, star apple, and garcinia, equal the strictly ornamental shrubs and small trees for landscape planting. And several of Florida's large trees are fruit producers, including the mango, sapodilla, jambolan plum, tamarind, and the lychee. No flowering tree is more colorful than a mango, lychee, or West Indian cherry when in full fruit.

Fruit producing trees and shrubs are sometimes used exclusively to landscape home grounds, and very effectively. The carissa and its several varieties are among south Florida's more popular ornamental shrubs. The cattley guava is widely used as a background plant and for screening. The Surinum cherry is one of the most commonly used shrubs for making hedges. The carambola, star apple, longan, and *Dovyalis abyssinica* are densely foliaged small trees recommended for use where a large screen or massive background is desirable. For unusual trees we have *Garcinia livingstonei,* the jackfruit, and the carob. Among the fruit producers are two palms, the coconut and the butia. And three of south Florida's most storm-resistant trees are fruit producers, the tamarind, jambolan plum, and sapodilla. If large tropical foliage is desired we have the banana and the monstera, two of the most tropical-like plants grown in Florida. If a ground cover in full sun is wanted we can turn to the pineapple, excellent for this purpose.

Thus it is possible to restrict landscape planting to the use of fruit producing plants and have as striking a result as it is possible to attain with primarily ornamental trees and shrubs. If indoor arrangement materials are wanted, tropical fruits equal any of the most colorful flowers.

Some knowledge of tropical fruit trees and shrubs is needed before we can use them effectively in a landscape plan however. We need to know a plant's ultimate size, form, leaf color, and texture, as well as the color of fruit; and we also need to know its cultural requirements, its hardiness to cold, and its tolerance to pruning. Not all tropical fruit trees will grow in south Florida. The breadfruit is likely to be badly injured by temperatures below 40 degrees, while it is sure to be frozen to the ground if the temperature drops to 32 degrees. The soursop (*Annona muricata*) is suitable for planting only in the warmer areas along the coast of south Florida and in the Florida Keys. This is a very effective evergreen tree, but it is likely to shed all its foliage if the temperature drops below 40 degrees.

A few trees, including the sugar apple, governor's plum, and white sapote, shed their leaves in winter and are likely to be without foliage for a few weeks. The common guava is objectionable to many persons because it drops its heavy crops of fruits to the ground, where they sour and give off the familiar penetrating guava odor. Other trees which are objectionable because of this habit are the jujube (*Zizyphus mauritiana*), jambolan plum, and egg fruit.

While most tropical fruit trees grow reasonably fast, a few are very slow. The jaboticaba may take eight years to get as tall as one's head, and may be ten to twelve years old before it begins fruiting. The grumnichama is another slow grower, especially in its young stage. It begins to bear about the fifth year. The sapodilla is fairly slow in its younger stage, but after it reaches five or six feet in height, growth tends to be much more rapid.

To plan, develop, and maintain the average kind of garden requires some knowledge of plants. To keep a superbly planted ornamental garden in good condition requires considerable knowledge of gardening. The same is true for the fruit garden. One really should be a good horticulturist and willing to read books on garden subjects. For some reason, those who go in for fruit production are likely to take the horticultural side more seriously than those who grow only ornamentals. Perhaps a fruit grower is more practical. Basically, there is no difference between the growing of ornamentals and the growing of fruit trees. To produce high quality, colorful flowers or fruits the plants must be well cared for. They must receive periodic applications of fertilizer, and be kept watered during droughts. They must be kept relatively free from insects and diseases.

Plants which produce large quantities of flowers or fruits are likely to develop deficiencies, especially if they are neglected. When an avocado or mango tree produces an unusually large crop, extra amounts of fertilizer should be applied to make up for the nutrients removed from the tree by the maturing fruits. If trees in heavy production are neglected they may suffer a set-back from which they may take two to three years to recover. Should a large healthy tree, such as a mango or avocado, fail to produce a crop of fruit in any year, heavy applications of fertilizer may not be needed. Only enough should be provided to keep the tree in healthy condition, unless greater growth is desired. Young, growing trees may be given several applications of fertilizer a year. It is the practice of avocado growers to apply small amounts of fertilizer to young trees monthly for the first year or two, then cut down applications to three or four a year. After a tree has reached maturity it is fertilized as fertilizer is needed.

It is especially helpful in Florida to provide fruit trees with large amounts of mulch, in order that satisfactory organic soil conditions may be developed. In soils low in organic matter, fruit trees are likely to suffer severely from mineral deficiencies, especially from copper, zinc, manganese, and iron. Such problems may be reduced considerably through the use of mulches, which add humus to the soil. Fertilizer is applied directly to the mulch, but care must be exercised here, because more fertilizer will be required than where mulch is not used. Microorganisms take up a part of the nitrogen for their work in breaking down the cellulose in the mulch. But the nitrogen is not lost. It is stored as in a bank, and is eventually returned to the soil. Through the combination of mulches and fertilizer an ideal soil situation is built up. Mulches also tend to cut down the activities of nematodes.

Mineral sprays to prevent mineral deficiencies should be applied regularly to young fruit trees. Many mature fruit trees, including citrus, lychee, carambola, jaboticaba, mango, avocado, and West Indian cherry, may require at least one application of nutritional spray a year. In early stages of planting, especially in limestone soil, young trees are benefited by nutritional spray two or three times a year, along with one or two applications of chelated iron. After the second year the application of nutritional spray should be cut to once a year. It is possible to build up a toxic level of these minerals in the soil if they are applied several times a year over a long period.

Design Rules the Same

The same rules apply to the design of the tropical fruit garden as the ornamental garden. Fruit collectors, tending to be more practical than the growers of ornamentals, often like to put their trees in neat rows, filling all spaces of the front and back yards with citrus, mangos, avocados, and other fruit trees. But row planting should be reserved for commercial plantations. Fruit producing shrubs or trees can, in many instances, be massed together for landscape effect, just as ornamentals are massed. It is true that by doing so we reduce production, but in the end we will be producing as many or more fruits than we can use anyway.

In an avocado grove, trees are planted twenty-five to thirty feet apart. In the back yard several varieties may be planted eight to ten feet apart. They will grow up into a mass, and production will be cut to some extent, but here again we will still produce as many avocados as we can use. Furthermore, by producing several varieties we gain an interesting tropical effect. Falling brown leaves may be left to lie on the ground beneath the trees in order to give contrast.

On a small city lot we cannot plant all of the fruit trees available. One or two mangos are about all such a lot will support. This large, spreading tree will require a great deal of space. If well grown, the lychee also makes a fairly large tree, with a spread of about thirty feet. If crowded, however, it will grow more upright in form than if permitted to spread out. Small to medium sized trees can be restricted to borders. The entire back yard should not be crowded with fruit trees, but an open area should be left for grass, as is recommended for any strictly ornamental planting.

Citrus trees, if trained to grow as shrubby specimens, with foliage extending to the ground, make excellent screens. But they must be planted far enough from a lot line to permit them to develop normal growth. If fruit trees over-grow the space set aside for them, with their branches invading neighbors' property, the neighbors should be invited to share the fruit, since fruit on over-hanging branches legally belongs to them anyway.

Improved varieties can be propagated only vegetatively, by budding, grafting, or airlayering. Growing these trees from seeds is usually disappointing because of the poor quality of fruit produced. Considerable information on tropical fruits, their culture and their uses can be found in libraries. Several books are in print, as well as bulletins published by the University of Florida. Research with tropical fruits is conducted at the Subtropical Experiment Station at Homestead, a branch of the University of Florida, and at the United States Plant Introduction Garden at Chapman Field. The annual proceedings of the Florida State Horticultural Society contain much valuable information about tropical fruits. The Rare Fruit Council is made up of members who are especially interested in the culture of tropical fruits.

Below is a list of tropical fruits which may be grown in south Florida. Those referred to as tender can be grown only in the protected parts of the peninsula and in the Keys.

ACHRAS ZAPOTA. Sapodilla. Spreading broadleaf evergreen tree to forty feet. Pulp of sweet apple-size brown fruit eaten fresh. Planted as shade tree. Seeds or grafting. Quite tender to cold when young. See *Trees and their Uses.*

AEGLE MARMELOS. Bael fruit. Small irregular tree, planted only as oddity. Scarcely worth planting for its fruit. Pulpy fruit, up to four inches in diameter, is covered with a woody, hard rind.

ALMOND. The true almond (*Prunus amygdalus*) does not grow in Florida. The tropical almond tree in Florida is *Terminalia catappa,* not related to the true almond. The nut of the tropical almond is eaten in the Orient, but is not eaten in Florida. Falling leaves present a problem. Seeds.

ANACARDIUM OCCIDENTALE. Cashew. Although this tender tree is the source of the cashew nut, it is planted in south Florida only as an oddity. The fleshy cashew apple may be eaten out of hand, but the nut, which grows from the base of the fruit, is covered by a thick shell which yields a caustic juice and can cause severe skin irritation. Seeds.

ANANAS COMOSUS. Pineapple. Ornamental plant to four feet, yielding edible pineapple in crown. Suitable for use as ground cover in sun or part shade. Propagated from suckers, as well as shoots which emerge from the base of the fruit. The crown of the fruit itself also may be planted. For home planting, plants should be started in five-inch clay pots and transplanted later to soil which has been fumigated to kill nematodes and weed seeds.

ANNONA CHERIMOLA. Cherimoya. Small, irregular tree which loses its leaves in winter. Not recommended because fruiting is never satisfactory. Cherimoya is a famous fruit in the highlands of Central and South America, but it does poorly at low elevation.

ANNONA hybrids. Atemoya. This cross between a cherimoya and a sugar apple does well in south Florida. The tree resembles the sugar apple in growth habit but the fruit is much like the cherimoya. Budding or grafting onto seedlings of custard apple, sugar apple, or native pond apple stock.

ANNONA MURICATA. Soursop. Small tree producing large irregular fruits up to five pounds, mainly on trunk and on large branches near trunk. The fruit is covered with soft spines. Pulp is used for making ice cream and drinks. The soursop is subject to cold injury and should be planted only in the warmer parts of south Florida and the keys. Seeds.

ANNONA RETICULATA. Custard apple. Small, irregular tree producing fruit of poor to fair quality. A poor investment both as a fruit and as an ornamental. Seeds.

ANNONA SQUAMOSA. Sugar apple. Small tree similar to custard apple, but producing superior fruits. Tree is bare of foliage for several weeks during winter. Like other annonas it is strictly for fruit collectors. Seeds. Superior varieties sometimes budded.

ANTIDESMA BUNIUS. A small, ornamental spreading evergreen tree, producing heavy clusters of red and black fruit. Fruit may be eaten out of hand. It also makes a fine red jelly, while the juice is used for making drinks. Superior fruiting varieties may be propagated by airlayering.

ARTOCARPUS ALTILIS. Breadfruit. Too tender to cold to be grown on mainland of Florida. Grown only to fruiting size in lower Florida Keys.

ARTOCARPUS INTEGRIFOLIA. Jackfruit. Although it can be grown in protected areas of south Florida, the jackfruit is seldom seen. Very large fruits, weighing up to twenty pounds, are produced along the trunk and larger branches. The yellowish, soft pulp is edible. Grown mainly as an oddity. The tree is quite tender to cold when young.

AVERRHOA CARAMBOLA. Carambola. A desirable medium sized tree, producing very juicy, yellow-orange fruit which is shaped like a star when cut cross-section. Sweet varieties may be eaten out of hand or in salads. Drinks can be made from the juice. Improved varieties may be propagated by grafting or airlayering.

AVOCADO. See *Persea americana*.

BANANA. See *Musa.*

BLIGHIA SAPIDA. Akee. Medium sized, ornamental evergreen tree, yielding three-celled capsule-shaped fruit about three inches long. The yellow fruit takes on a bright red blush as it approaches maturity. After maturity the fruit splits and exposes three black seeds partly embedded in a white, edible aril. The arils are widely used in the West Indies, and especially in Jamaica, as food. They are often fried in butter, but may be chopped up and used in salads. The akee arils are said to be poisonous before the fruit opens, and they are also said to become poisonous again if allowed to age and become soft. Planted mainly as an ornamental shade tree.

BUTIA CAPITATA. See *Palms.*

CALOCARPUM SAPOTA. Sapota. A fairly large tree which has been rarely planted in south Florida. The russet-brown fruits grow to five inches in length. The reddish plup is eaten fresh. Young trees are tender to cold.

CARICA PAPAYA. A large herb-like, soft-stemmed plant reaching ten feet or more in height in a single year, if grown under favorable conditions. Unfortunately, a virus disease makes it difficult to grow worthwhile papayas in Florida. Stunted, yellow plants affected with virus should be removed immediately from any planting of papayas to prevent spread of disease. Melon-size fruits mature in a year from planting time. Grown from seeds. Male, female, and bisexual plants are produced. One or more male plants should be preserved to insure pollination. Male flowers are borne on stems up to a foot or more in length. Female and bisexual flowers are borne on short stems.

CARISSA GRANDIFLORA. Several varieties of this large shrub widely used in landscape plantings. See *Shrubs.* Bright red fruits, up to two inches long, are used mainly for making jelly.

CASHEW. See *Anacardium occidentale.*

CASIMIROA EDULIS. White sapote. A small irregular tree producing large quantities of green to greenish-yellow fruits that to some resemble apples at a distance. Creamy, yellowish flesh is variable in quality. Superior varieties may be propagated by budding or grafting.

CERATONIA SILIQUA. Carob. St. John's bread. Small tree producing edible bean-shaped pods. Seldom planted except as an oddity.

CEYLON GOOSEBERRY. See *Dovyalis hebecarpa.*

CHRYSOPHYLLUM CAINITO. Star apple. Medium size tree to thirty feet, with striking foliage which is deep green above and copper colored underneath. Trees are tender to cold and can be grown to maturity only in warmer areas of south Florida. Fruits, to four inches in diameter, are variable in quality. Grown principally as an ornamental rather than for fruit production. Seeds.

CITRUS. See special chapter, *Dooryard Citrus.*

CLAUSENA LANSIUM. Wampi. A small tree producing panicles of acid-sweet fruit about one inch in diameter. Not widely planted. Seeds.

COCCOLOBA UVIFERA. Seagrape. Although widely planted as an ornamental shrub or tree, the seagrape is often listed under fruit trees because of its heavy production of grape-like clusters of fruits. A distinctive jelly is made of the fruits. Trees of male and female sexes are required for fruit production, which explains why so many dooryard seagrape trees never yield. (See Trees). Airlayering.

COCONUT. See *Palms.*

COFFEA ARABICA. Coffee. A small shrub grown mainly for its ornamental white flowers and bright red fruits in south Florida. Not grown commercially in Florida.

DIOSPYROS DISCOLOR. A small ornamental tree bearing fruit to three inches in diameter which is covered with reddish brown fuzz. Grown mainly for ornament. Seeds.

DIOSPYROS EBENASTER. Black sapote. A small evergreen tree which is ornamental as well as productive. The fruit is two to five inches in diameter and dark olive green at maturity. The soft, chocolate-brown flesh is sweet but insipid and may be improved by adding orange or lemon juice. Seeds.

DOVYALIS ABYSSINICA X DOVYALIS HEBECARPA. A small hybrid tree distributed by the U. S. Plant Introduction Garden. This tree is more vigorous than either parent. Produces large quantities of brown-skinned fruit with a yellow flesh. The flavor suggests fresh apricot. Airlayering.

ELAEAGNUS PHILIPPENSIS. Lingaro. Climbing or sprawling shrub, attractive as an ornamental as well as a producer of edible red fruits about the size of a small olive. The flavor is tart but sweet, and the fruits may be eaten out of hand or made into a highly colorful jelly. Seeds or cuttings.

ERIOBOTRYA JAPONICA. Loquat. A symmetrical small evergreen tree grown as much for ornament as for its production of orange-yellow fruit. The fruit, however, is of very high quality on selected trees and may be eaten out of hand or made into a jelly. It averages about one and one-half inches in length. Grafted trees are much more dependable than seedlings.

EUGENIA AGGREGATA. Cherry of the Rio Grande. A desirable ornamental shrub, producing small fruits which turn from orange red to purple red while ripening. Rare in south Florida. Seeds.

EUGENIA BRASILIENSIS. Eugenia dombeyi. Grumichama. A small evergreen tree producing large quantities of purplish-black fruits nearly one inch in diameter, of agreeable flavor. A very desirable ornamental tree. Growth is slow. This tree, like all eugenias, requires large amounts of water during fruit development. Seeds.

EUGENIA LUSCHNATHIANA. Pitomba. A small shrubby ornamental tree or large shrub, producing numerous orange-yellow fruit about one inch long. Flavor agreeable but aromatic. Seeds.

EUGENIA UNIFLORA. Surinam cherry. Florida cherry. A widely grown ornamental shrub in south Florida, producing numerous red to black fruit up to one and one-half inches in diameter. Fruits of superior varieties are very good when eaten fresh. Fruits of many seedlings may have unpleasant aromatic flavor. Seeds.

EUPHORBIA LONGANA. Longan. A small, round evergreen tree related to the lychee, and producing heavy clusters of reddish brown fruit about one-inch in diameter. The fruits are eaten fresh. A useful ornamental tree, and the fruit is fairly good. Airlayering or seeds.

FICUS CARICA. Fig. Easily grown in South Florida if protected from nematodes and leaf rust. Heavy mulching aids in keeping down nematodes, while monthly spraying with copper fungicide may be necessary to control leaf rust. Celeste is only variety which thrives here. Cuttings.

FLACOURTIA INDICA. Governor's plum. A small spreading tree or large shrub, producing large quantities of purple or black fruits up to one inch in diameter on female trees. A male tree is needed for pollination. The flavor of fruits on different trees vary considerably, and only the better varieties should be propagated, by airlayering or by rooting cuttings. While the governor's plum is a widely planted ornamental, it tends to lose its leaves for a very short while in the spring.

GARCINIA LIVINGSTONEI. Imbie. A small very irregular tree, very unusual in its growth. The orange colored fruits, something over an inch in diameter, are eaten out of hand. A striking ornamental if a small tree is desired, but messy when in fruit. Seeds.

GARCINIA MANGOSTANA. Mangosteen. A small evergreen tree which does not appear to be well adapted to south Florida. It seems to prefer not only a warmer climate but an area where there is heavy rainfall throughout the year. Fairly large trees have been grown in the warmer parts of south Florida, but while some have flowered, no production of fruit has been reported.

LITCHI CHINENSIS. Lychee. A medium sized, round evergreen ornamental tree, producing large quantities of bright red fruits nearly two inches in length. Fruit, ripening in June, is excellent fresh or frozen. Tree is fairly hardy to cold, but requires considerable cultural care in its younger stage. When planted in limestone soil, one or more applications of neutral or chelated iron are needed each year to prevent the foliage from becoming yellow.

MACADAMIA TERNIFOLIA. Macadamia nut. A medium sized evergreen tree with spiney foliage reminding one of American holly. This is the only tropical nut producing tree planted in Florida. Small, very hard nut up to one inch in diameter, must be cracked with vise or with vise-grip pliers. Tree has not been a heavy producer in south Florida.

MALPIGHIA GLABRA. West Indian cherry. Barbados cherry. Acerola. A large spreading shrub or small tree, producing several crops a year of bright red fruits one inch in diameter or less. Several crops may be produced between April and late fall. Fruit varies on different varieties, from very sour to sweet. This fruit gained wide attention when found to contain more ascorbic acid than any other fruit of the same size. Because the tree is subject to damage from soil nematodes, it should be kept well mulched to help keep down infestation by these pests. Superior plants may be propagated by airlayering or rooting cuttings.

MAMMEA AMERICANA. Mammee apple. Mamey. An ornamental medium sized broadleaf evergreen tree producing large russet-colored fruits up to six inches in diameter. The orange or reddish flesh has an agreeable acid-sweet flavor. It may be eaten out of hand or served with sugar and cream. Some trees mature their fruits throughout the summer. In south Florida propagation is limited to the planting of the large seeds, but small plants are very tender to cold and also have difficulty getting started in limestone soil. A fairly large mamey tree, said to have been planted soon after the turn of the century, is one of the most striking specimens on the grounds of the Fairchild Tropical Garden.

MANGIFERA INDICA. Mango. One of the oldest cultivated and most renowned of all fruits, and a very popular dooryard tree throughout the tropics, including south Florida. Dozens of mango varieties are planted in Florida, as well as types known as West Indian, Saigon, and Indian. The West Indian, which includes the turpentine, is the poorest of all mangos.

The mango season begins usually in June with the ripening of the Haden, Zill, Irwin, Carrie, and Saigon, and ends in September with the ripening of the Sensation, Keitt, and Brooks. Midseason varieties, ripening in July and August, are Edward, Lippins, Fascell, Springfels, and Kent. Taste preferences vary greatly and only by tasting a mango can one judge its quality. Most widely planted in dooryards are the Haden, Zill, Irwin, Kent, Sensation, and Keitt. The Carrie and the Edward have been acclaimed as two of the finest flavored mangos grown in Florida, but neither is of any value for commercial planting. The Carrie lacks external color, being only yellow-green when ripe, while the Edward is a shy producer. Carrie, Edward, and Florigon are recommended for home planting because of disease resistance.

Mango trees should be planted where they have ample room to grow, and, since they must be sprayed from time to time, it is best to plant them some distance from a house or a patio. Anthracnose, a fungus disease, is the greatest enemy of the mango. It attacks leaves, flowers, and fruits. A fungicide should be applied at the time individual flowers are beginning to open on their panicles, in winter, and two or three times thereafter as the fruit begins to form and while it is developing.

A mango tree which is large enough to produce fruit should not be irrigated from the end of the rainy season until it has begun to set its fruit. Then irrigation is not necessary unless the soil is excessively dry. Over-irrigation can discourage bloom and also cause the development of poor fruit.

Some persons are allergic to the mango and cannot eat the fruit without taking some precautions against coming in contact with its skin. Mango poisoning, or mango dermatitis as it also is called, shows up as a rash on the skin, similar to that caused by poison ivy. Sensitive persons develop rash about the mouth and on the hands. If the fruit is peeled by a non-sensitive person, and then washed, the sensitive person sometimes may eat it without harm. But the knife used for peeling the fruit must be washed if it also is to be used for cutting up the fruit. Some persons are so highly sensitive to the mango that they cannot come in contact with the foliage or pollen without developing a skin rash.

Only budded varieties should be propagated, as seedling trees seldom produce worthwhile fruit.

MELICOCCA BIJUGA. Spanish lime. A fairly slow growing tree reaching large size, bearing clusters of juicy, acid fruits about one inch in diameter. The thick skin is green. This tree is particularly popular at Key West, where there are many large trees. It is less widely grown on the mainland, where it does not seem to fruit so well as at the southernmost Florida city. Seeds.

MONSTERA DELICIOSA. A climbing aroid with very large, incised leaves more than two feet in length and nearly as wide. Widely grown as an ornamental outdoors in south Florida and indoors in the north. Can be grown as a sprawling shrub or as a climber on large trees. The fruit, the size and shape of a slender ear of corn, develops from a large calla-like flower. It takes fourteen months for the fruit to mature. Ripening is indicated by a slight yellowing at the base of the fruit, and the development of a strong pineapple-banana like aroma. A fruit may be harvested when the first scales begin to loosen. The edible part is covered by these scales. Fruits produced in Florida seldom have many seeds, unless the flowers are hand pollinated. The monstera is more widely grown for ornament than for fruit production.

MUNTINGIA CALABURA. Capulin. A very rapid growing small tree, often yielding fruits within two years after planting. The fruit is a yellow or red berry about one-half inch in diameter with a sweet, juicy pulp containing many small seeds. The fruits may be eaten out of hand or made into jam. A good tree where there are children. Seeds.

MUSA. Banana. While there are hundreds of banana varieties, only a few have been introduced to Florida. Most widely cultivated is the Cavendish or Chinese dwarf. Other common varieties are Lady-Finger, and the Orinoco or Horse banana. The Orinoco is the hardiest of all banana varieties, being grown in protected places as far north as Jacksonville. The fruit is very poor unless permitted to ripen until the flesh is soft. This variety also can be cooked as a plantain, as is done in the highlands of the tropics where other banana varieties do not thrive. Plants should be fertilized three or four times a year and kept mulched with partly decomposed plant materials. Bananas may be harvested when mature-green or after they have ripened, which is indicated by a slight yellowing of the skin. Maturity is reached when the individual fruit has lost its slender, angular form and is well rounded when cut in cross-section. A banana stalk produces but once. Additional crops are produced by suckers which come up from the base of the older plant. Bananas are propagated by dividing old clumps and chopping up the rhizomes, or by transplanting individual suckers, which may be removed from older plants by the aid of a sharp, flat spade.

MYRCIARIA CAULIFLORA. Jaboticaba. A very slow growing shrubby tree related to the eugenia, producing its fruits singly or in small clusters along the trunk or branches. In limestone soil young plants may require periodic applications of chelated iron to prevent leaf yellowing. Keeping the plants mulched with half-decayed organic matter, such as partly composted leaves and grass clippings, usually helps this tree become adapted to limestone conditions. Several crops of fruits a year may be produced by older trees, if given plenty of water and fertilizer. The fruit, up to one inch in diameter, is best eaten fresh, directly from the tree. Seeds.

OLEA EUROPAEA. Olive. Grown rarely as an ornamental tree in south Florida, where it does not flower or fruit. See *Trees.*

PAPAYA. See *Carica Papaya.*

PASSIFLORA. Passion flower. Several species of the passion flower are grown in south Florida, usually more for ornament than for fruit production. Most productive in Florida is *Passiflora edulis,* commonly known as the purple granadilla. Its most serious problem is a crown rot, caused by a fungus. The purple fruit is about two inches in diameter. Less widely planted is *Passiflora quadrangularis,* or the giant granadilla. The edible part of the passion flower fruit is a sweet, juicy pulp in which the seeds are embedded. One species of passion flower, the maypop, grows from north Florida to Virginia and west to Texas. Most of the other species are tropical. The passion flower makes an aggressive, tall growing vine when planted in soil free of nematodes. Its planting is limited in south Florida because it is highly susceptible to nematode attack. Seeds or layering.

PEACH. One type, the Ceylon, thrives in south Florida. The fruit, although small, is colorful, free-stone, and of good flavor. It is a worthwhile dooryard fruit tree, but may seem out of place in a tropical garden. No superior variety of the Ceylon type has been selected. Seeds.

PERSEA AMERICANA. Avocado. This highly productive fruit tree is well adapted to south Florida and is easily fitted into the small lot landscape plan. If several varieties are planted the harvest season can be extended from the end of July to the first of March. Here are four recommended varieties: Pollock, late July to September; Waldin, September to early December; Booth 7, November to January, and Choquette, February, and March. Like other fruit trees, the avocado is responsive to good culture. First and second year trees must be protected from cold.

PISTACHIO. The true pistachio (*Pistacia vera*) does not grow in Florida. The small tree sometimes grown here as a pistachio is really the physic nut (*Jatropha curcas*), which is a cousin of the castor bean plant and not even distantly related to the true pistachio. Children and adults have been hospitalized after eating the oily nuts from the physic nut tree.

POUTERIA CAMPECHIANA. *Lucuma nervosa.* Canistel. Egg fruit tree. A small evergreen tree producing fruits up to five inches long and three inches in diameter. The orange-yellow flesh has the texture, and a touch of the flavor of a baked mealy sweet potato. Varieties vary greatly in flavor. Some seedlings have a musky, unpleasant flavor. Fruits mature in late fall and early winter. Seeds or grafting.

PSIDIUM CATTLEIANUM. Cattley guava. A hardy, shrubby tree, widely used in landscape planting, especially for screening and for background. The red or reddish-purple fruits, a little over one inch in diameter, may be eaten fresh or used for jelly making. Airlayering. Seeds commonly used by nurserymen. See *Shrubs.*

PSIDIUM GUAJAVA. Common guava. A small, sprawling tree whose fruit is widely used for the making of jelly. Improved varieties have been developed by the Subtropical Experiment Station which produce large, sweet fruits which may be eaten fresh or used for making guava shells. Only sour variety should be planted if fruit is wanted for jelly making. The tree is thrifty and can be pruned to take almost any form desired. Airlayering.

PUNICA GRANATUM. Pomegranate. Planted only for ornament in south Florida where the summers are too humid for the production of high quality fruits. Very susceptible to leaf spot caused by a fungus.

SPONDIAS MOMBIN. Hog plum. A vigorous spreading tree in south Florida, reaching thirty feet or more. Yields large clusters of yellow fruits up to one and one-half inches in diameter. Most of the Florida seedlings are inferior in quality. The season is in late summer and early fall. Seeds.

SYZYGIUM CUMINI. Jambolan plum. More widely planted for shade, this tree bears large quantities of deep purple fruits up to two inches long, ripening in late spring and early summer. Although the fruits are edible they are seldom used. In fact, the fruits are often looked upon as objectionable because large quantities of them lie on the ground and sour, and produce a wine-like but unpleasant odor. Seeds.

SYZYGIUM JAMBOS. Rose apple. A small spreading tree producing fruits up to two inches in diameter in late spring. Fruit is insipid, having something of the flavor of rose water. Some varieties make excellent jelly and jam. Not widely planted. Seeds.

TAMARINDUS INDICA. Tamarind. A large very slow growing tree planted mainly because it is so resistant to high wind, than for the quality of its fruits. The edible part is borne in a brown pod up to six inches in length. The pulp is used for making preserves and chutney. It has a rather pleasant acid-sweet flavor. Seeds or airlayering. See *Trees.*

ZIZYPHUS MAURITIANA. Jujube. This medium sized tree, with willowy branches, produces large quantities of fruit slightly more than one inch in length. Most persons find the flavor very poor. The local jujube is often confused with the famous Chinese jujube (*Zizyphus jujuba*) which produces a fruit much superior to that of the common *Zizyphus mauritiana.* The Chinese jujube is unthrifty in Florida, on account of the humid conditions, and is seldom planted.

Tasty citrus can be grown in dooryard if given good care

Tropics offers gardener wide choice of interesting plants

Chapter 10

DOORYARD CITRUS

Citrus is the most commonly planted dooryard tree in Florida, both for ornament and for fruit production. Few other trees equal it in serving this dual purpose. It is worthy of its keep every month of the year. In late winter the tree is covered with fragrant, waxy white bloom. From late fall until spring the branches are laden with colorful ripening fruits. Moreover, a well grown citrus tree, with its dark green foliage, equals the finest laurel, even when bearing neither bright fruits nor fragrant flowers.

Newcomers to Florida are especially attracted to citrus and often plant several varieties of oranges and grapefruit, with the hope that they soon will be able to harvest their own fruits from dooryard trees. Their hope is usually blighted, though, because not only do the trees generally grow poorly, but the quality of the fruits is much inferior to that which can be bought in the grocery store. It is evident that the homeowner needs to know more about the growing of citrus, and about the numerous varieties which can be grown in Florida.

It seems hardly worthwhile to give valuable space in a small lot to a common orange tree, such as a Pineapple or Valencia. Millions of boxes of these two well known oranges are produced each year in Florida, and good quality fruits can be purchased much cheaper than the homeowner can produce them. Why not grow varieties that are rarely seen in stores? Some of these are of exceptionally high quality when the trees are properly maintained – and all citrus must be well maintained if the fruits are to be worthwhile. Several varieties are available for dooryard planting which are little grown commercially. How often do we see Minneola tangelos or Rangpur limes for sale at the grocery store? Why grow Persian limes when key limes are easier to grow and are superior in quality? Some of the best citrus varieties are found among the tangelos, a cross between a grapefruit and tangerine.

If an orange is wanted the Temple is hard to beat. It is easy to peel and is probably the best of all oranges for picking directly from the tree and eating out of hand. And this is a major part of the pleasure in producing dooryard fruits. Where one plants an extensive dooryard citrus collection, the Hamlin orange may be added simply because it matures earlier than most early citrus fruits, being ready to pick by mid-November. The flavor improves until Christmas, when the Hamlin is at its very best. Hamlin fruits will remain on the tree through January.

Among the finest of ornamentals are the Cleopatra tangerine and the calamondin. Both can be trained as low, spreading shrubs. They will hold their colorful fruits throughout the winter and far into spring. The Cleopatra fruit is of fairly good quality but is the smallest of the tangerines. The calamondin, very sour, is popular for preparing drinks or for making marmalade.

Selecting the Right Rootstock

In general, citrus trees do not become nearly so large in southeast Florida's limestone soil as they do in the major citrus growing areas of central Florida. Good culture not only is important, it is a vital necessity. Unless trees are budded on the right kind of rootstock and receive proper attention, the fruits generally may be of very poor quality – thick skinned, pulpy, and juiceless.

Citrus is not an easy tree to grow. It is subject to many nutritional troubles, is host to many insects, and is attacked by many diseases. It has been said appropriately that anyone who can grow high quality citrus in Dade County can grow anything anywhere in the world. This does not apply to those who are fortunate enough to have excellent hammock soil, in which citrus does well, even in limestone.

First of all, one should get full information about the different citrus varieties and learn something about the problems of citrus growing before planting the first tree. It is nearly always sold as a budded tree. Several kinds of root stocks are used, including rough lemon, sour orange, trifoliate orange, and Cleopatra tangerine. South Florida nurserymen usually bud citrus on rough lemon root stock, because it makes a vigorous tree. It is not the best, by far. Trees budded to Cleopatra root stock produce much superior fruit. However, growth of young plants is exceedingly slow in limestone soil, and for this reason nurserymen do not like to use Cleopatra. They have to ask higher prices for trees, to which the buyer objects. But the higher cost is worth the difference. Sour orange also makes an excellent root stock, but is planted mainly in low areas where the water table is high, and especially in areas subject to periodic flooding. Most of the citrus trees in the Indian River section of Florida are budded onto sour orange root stock, which is one major reason for the high quality fruit produced there. Sour orange stock is not recommended for trees to be planted in Dade County, except possibly in low areas where the water table remains fairly high.

Most ornamental trees and shrubs can be kept in fair condition by ordinary practices of using common garden fertilizer and by irrigating during drought. Citrus requires more than this. It requires, first, a high quality fertilizer containing a high percentage of organic nitrogen. In central Florida, where large amounts of citrus fertilizer are used, magnesium sulphate is added. Special citrus fertilizer may not be obtainable in south Florida, but magnesium sulphate is inexpensive and is sold by all drug stores, if unavailable in garden supply stores.

Fertilizer is applied three or four times a year. Applications may be made in January, May, August, and October. If only three are made, pass up the August application. Because of the many different fertilizer analyses, it is difficult to suggest an accurate guide for applying fertilizer to individual trees. Where a 6-6-6 analysis is used, (each figure represents the percentage of nitrogen, phosphorous, and potash) one-half to one pound of fertilizer may be applied for each year of the tree's age. A tree five years old would get two and a half to five pounds per application, depending on its size. As with other trees, fertilizer should be applied evenly on the soil beneath the tree canopy, but never close to the trunk itself.

Lawn grass should be kept cut back from citrus trees, in a circle close to the outer edge of the leaf canopy. Grass not only robs a citrus tree of fertilizer, but of water during drought. Mulching is important. Organic matter, such as grass cuttings, leaves, or composted material, should be applied over the soil beneath a citrus tree from time to time. Such material should be kept away from the trunk. Because citrus is subject to fungus attack at the soil level, this area should be kept clean of all debris in order to insure good aeration.

In addition to fertilizer and mulching, citrus must have at least one annual application of a mineral spray containing copper, zinc, and manganese. These materials may be obtained in neutral form and may be applied without fear of burning the foliage. January is usually the time for the annual nutritional spray, but if trees are in poor condition and produce poor quality fruit, a second application in the summer may be necessary. Trees growing in limestone soil also are subject to iron deficiency and may be helped to a large extent by applications of chelated iron.

Insects, including scales, aphids and mites, are serious pests of citrus. Scales, usually seen on the leaves, may be controlled by spraying with malathion or oil emulsion. Malathion or nicotine sulphate will control aphids, which attack new growth and cause it to become distorted. Rust mites which cause russeting of fruits and foliage, with consequent foliage dropping, can be controlled by spraying with zineb or by dusting or spraying with sulphur every two months from the time of fruit setting until fall. Zineb also controls a serious leaf disease, "greasy spot."

Where large groves are planted it is necessary to set trees in orchard rows for efficiency in culture and harvest. Such planting is hardly desirable in the dooryard, and trees should be fitted into the landscape according to the best design rather than set up in rows like soldiers. Trees can be used for informal or sheared hedges and screens and to some extent for background planting. The smaller members of the citrus group can be used to a limited extent for foundation planting, where it is desirable to have large shrubby plants at corners or to cover large bare walls. In general, however, citrus should not be planted so close to the house that spraying is difficult without getting chemicals on wall.

Avoid Sprinkler System

It is questionable whether good citrus can be produced where lawn sprinklers are permitted to play on the trees several days each week, keeping both the trees and the soil wet. If budded onto sour orange root stock such trees probably would have a much better chance of producing high quality fruit than trees budded onto rough lemon root stock. Ordinarily such citrus is low in sugars and acid, and it is likely to be pithy. Perhaps a citrus planting should be put in an area where watering can be controlled. If an established tree is thoroughly watered every three or four weeks during the driest part of the year that should be sufficient.

A word of caution should be offered against the planting of a citrus tree too deeply. If the bud union is covered with soil it can be fatal. The tree is likely to grow poorly, if at all, and eventually fungus is sure to get started in the soil-covered wood above the bud, resulting in death to the tree. Where the ground is very flat, it may be a good idea to plant a citrus tree on a very slight mound, perhaps two or three inches above the rest of the area. The mound should be made of good soil and should be five or six feet in diameter. In high, sandy areas which are quite well drained, this precaution is not important.

Below is a list of the citrus species which may be planted in south Florida, along with the better varieties of each. The method of listing may at first seem confusing to many persons, but it really is the only way to list citrus and keep the groupings from becoming even more confusing. For instance, we have three species of oranges, under which there are several varieties. The sour orange, isn't just a different variety from the sweet orange. These fruits are different species. We know of no way to do anything well without first learning something about the subject, and this is especially true with citrus. Unless the gardener is willing to learn something about citrus – its species and varieties, its rootstocks, its nutrition problems, as well as its pest enemies, he should not attempt to grow it. For he will harvest more disappointments than good fruits. Unless he is willing to learn, he should stick with the easy-to-grow ornamental shrubs and trees.

CITRUS AURANTIFOLIA. Lime. This group contains several distinct varieties, each noted for its special acid flavor. Trees are generally shrubby but irregular in form, and some varieties are armed with sharp spines. Only one variety, the Persian, is grown commercially in Florida, mainly in Dade County. Others are dooryard trees, of which the key lime is the most popular. Planted less widely is the Rangpur lime, believed to be a natural hybrid between a lime and a tangerine. The fruit resembles a tangerine in every respect but flavor, which is distinctly that of an acid lime. Another hybrid, the Otaheite orange, believed to be a cross between a lime and a tangerine, has a rather flat, sweet flavor. It is sometimes grown as an indoor ornamental plant for the northern trade, but is seldom propagated for dooryard planting in Florida. Limes mature their major crops in late spring and summer, but older trees may produce a few fruits all year.

CITRUS AURANTIUM. Sour orange. Seville orange. Bitter orange. Although used mainly as root stock for the growing of citrus in low, moist areas subject to flooding, the sour orange was probably the first citrus to be introduced to Florida. Indians scattered the seeds throughout the hammocks of central and south Florida, and large numbers of these old wild orange trees still stand today. The laurel-like trees are quite ornamental, and the orange colored fruits, which persist for several months during the winter and spring, add to the beauty of this tree. Although the fruit resembles a sweet orange in color and texture, the flavor is so highly acid and bitter that the fruit cannot be eaten out of hand. It may be used like a lime or for making a superior type of marmalade. In Spain this fruit is known as the Seville orange. Thousands of acres are grown for sale in England, where the fruit is made into preserves. The sour orange is usually propagated from seeds.

CITRUS GRANDIS. *Citrus maxima.* Shaddock. Pummelo. This is the largest of all citrus fruits, sometimes reaching twelve to fifteen pounds. The rind is very thick and the edible part is coarse and resembles the flavor of a grapefruit, but is inferior in quality. In south Florida the tree is usually small and scraggly and can be considered only in the class of oddities.

CITRUS LIMON. Lemon. Prior to the Florida freeze of 1894-95, lemon growing was an important Florida industry. Commercial production never made a come-back and eventually California took over the lemon industry. Several varieties are grown as dooryard trees in Florida, including Villafranca, Everbearing, and Meyer, all of which are good quality fruits. Eureka, the foremost lemon grown in California, is not widely planted in Florida, being susceptible to scab disease. The Ponderosa, largest of the lemons, reaching nearly five inches in length, is grown mainly as an oddity. Another variety, the rough lemon, has been the most commonly used root stock for the budding of Florida citrus during the last half century. It is especially used in central Florida and in Dade County. However, in Dade County it is much inferior to Cleopatra tangerine as a root stock. Citrus budded on to rough lemon should not to be planted in limestone soil if it is possible to obtain trees budded on Cleopatra.

CITRUS MITIS. Calamondin. A small, well formed tree producing numbers of orange colored fruits somewhat smaller than a tangerine. It is believed to be a hybrid between a kumquat and a tangerine. Too acid to be eaten out of hand, the fruit is widely used for mixing drinks and for making preserves. It is widely planted as a small ornamental tree or shrub. It can be trained by pruning to take a low sprawling form or a tree-like form.

CITRUS NOBILIS. King orange. Although this tree is seldom planted because of its open, scraggly shape, the large, rough, tangerine-like fruits are excellent in flavor. The fruits mature in winter.

CITRUS PARADISI. Grapefruit. There are several varieties of this fruit, maturing between November and February. The tree is somewhat larger and more spreading than the orange tree, but is still likely to be relatively small when grown in the shallow limestone soil of southeast Florida. The quality of the fruit grown in this area is seldom much to brag about when produced under average dooryard cultural conditions. The earliest maturing variety is Duncan, which may be eaten after the first of November, but improves in flavor until the middle of December. Marsh Seedless, which matures in January, may hang on the tree through May. It is the most widely planted grapefruit variety, because of its high quality. Foster, Ruby, and Thompson are pink varieties, reaching maturity in January. Confusion surrounds the pink grapefruit varieties, and nurserymen may simply sell them all under the name of "pink grapefruit," not knowing which is represented.

CITRUS RETICULATA. Tangerine. Mandarin orange. In this group are some of the most striking trees among the citrus varieties. These are small to medium sized, with deep green, laurel-like foliage. The fruits are highly colorful and very decorative. Although there are several varieties of tangerines, only one, Dancy, is widely planted, either commercially or as a dooryard tree. Cleopatra produces a much smaller, more irregular fruit than the Dancy tangerine. Although the Cleopatra is widely used as a citrus root stock in limestone soil, it is often described as the most worthy ornamental of all the citrus. Its brightly colored fruits hang on the tree for months. The tree is symmetrical and the foliage is very dark green and dense. Botanists sometimes include the Temple orange with the tangerine because it is believed to be a natural cross between a tangerine and a sweet orange. Another variety, the Satsuma, is widely grown in north Florida and also in southern Alabama and Mississippi. It is seldom planted in south Florida.

CITRUS RETICULATA X CITRUS PARADISI. Tangelo. The tangelo, of which several varieties are planted, was developed from a cross between a tangerine and a grapefruit, by agricultural department scientists soon after the turn of the century. The fruits have the distinctive, blended flavor of the grapefruit and the tangerine. In some varieties the grapefruit flavor predominates, and others the tangerine. Those in which the tangerine predominates are the ones most widely planted. The tangelos are about the size of a sweet orange. Most popular is the Minneola. It has a distinctive character, a short beak at the stem end. It is a cross between a Bowen grapefruit and a Dancy tangerine. The fruit color is that of a tangerine and the flavor is excellent. Almost as widely planted as the Minneola is the Orlando tangelo, but Thornton, Nokatee, and Seminole are all worthwhile varieties. The tangelos begin maturing in December and the fruit will hang on the trees until April. They need to be planted near other citrus, especially the sweet orange, to insure pollination and fruit setting.

CITRUS SINENSIS. Sweet orange. This is the most widely planted of citrus. Hundreds of varieties have been selected and named but only a few are planted commercially. Generally planted commercially in Florida are the Hamlin, maturing in October and November; Pineapple, January and February, and Valencia, March to June. The Washington Navel orange, a popular commercial variety in California, is not widely grown in Florida. Several other navel oranges are used in Florida, including Egyptian, Summerfield, and Surprise. None of these is planted commercially except for roadside sale or for special gift packages. The navel oranges should be budded on Cleopatra rootstock for planting in limestone soil. The trees require top cultural conditions if they are to produce worthwhile fruit.

The Temple, a hybrid between an orange and a tangerine, is often placed by botanists in the category with the tangerine, or Mandarin oranges. It is, however, looked upon by most persons as an orange and it bears the name "orange." It is a superior fruit when grown under good cultural conditions. It should be propagated on Cleopatra stock, except when it is to be planted in low, moist areas, in which case it may be propagated on sour orange stock. The tree resembles the tangerine tree in appearance. The season runs from about Christmas to April. Like the tangerine, the Temple orange makes an excellent dooryard fruit tree. It not only is highly ornamental but is simple to eat out of hand because it is easy to peel, a very desirable quality for dooryard fruit.

An orange which is seldom planted commercially and is seen in few dooryard collections is the blood orange. Ruby is considered to be the best. The fine-grained flesh is blood red when ripe. Unless budded on to Cleopatra it is not likely to produce excellent fruits in limestone soil. There are many other varieties of oranges, including Parson Brown, an early variety, and Lue Gim Gong, a late variety sometimes planted in the dooryard grove because it hangs on the tree into early summer. These two have no particular distinction so far as quality is concerned and hardly seem worthy of planting in the home garden when oranges of similar character are available at the grocery store.

FORTUNELLA SPECIES. Kumquat. Although this small shrub belongs to the citrus group, it has been placed in a genus by itself – *Fortunella*. Four species are grown in Florida, and anyone deeply interested in citrus is invited to check the proper references for further information. The variety Nagami (*Fortunella margarita*) is the most widely planted kumquat, and the shape of the fruit, like that of an olive, is usually associated with the kumquat. The Marumi variety (*Fortunella japonica*) also makes an excellent ornamental and produces high quality but round fruits.

This by no means exhausts the list of cultivated citrus. Occasionally we see cultivated several kinds of what to most of us may be strange-looking citrus fruits, including the citron (*Citrus medica*), limequats, citranges, and even citrangequats. The first mentioned, the citron, is grown commercially in the Mediterranean areas where the rind is cured and exported for use in confection manufacture. The others are various hybrids, drawing interest only from collectors and scientists doing research with citrus.

Grass is kept at a minimum in this Coconut Grove garden

Easy-to-maintain centipede is used in this large area

Chapter 11

LAWN GRASSES AND THEIR CARE

Trees and shrubbery provide form, texture and color in the garden; the lawn provides space and helps to give dimension. Without space we have a jumbled planting, lacking in order and in character. Open place should be carefully planned when the garden is designed. It is necessary not only for the free movement of humans, but also for the circulation of air and the entrance of sunlight. Space also serves another purpose. It is like the open areas one finds in a museum. Imagine the clutter and confusion in a museum where there was no thought given to openness, where objects were so closely packed that the viewer had no opportunity to see works of art except at extreme short range. The greatest collection of art in the world would have no interest for us in such disorganized conditions. The same is true in a garden where one can not back off and view plants from a distance.

Spaciousness can be attained by confining the planting of trees and shrubbery about the borders of a garden and leaving the center open. Except for the possible planting of one or two shade trees in appropriate locations, the center of the garden should not be cluttered with any kinds of plants. Most people prefer to cover the ground with grass, or such materials as gravel, paving, or even wood chips. Grass is by far the most popular. And it is the ambition of every one who has a lawn to find a kind of grass which remains green all year, requires little mowing or watering, and is resistant to insects and diseases.

There is no perfect grass, however, and so far no magic chemicals have been developed which control all insects and diseases and keep out all weeds. The quality of a lawn is dependent chiefly on the person who maintains it. A lawn which is half-starved part of the year, which is mowed infrequently, and which is also improperly watered is sure to be a problem. The poorest lawn will require some effort, no matter how relaxed one may feel about caring for it. Even the weeds must be cut from time to time. On the other hand, there is virtually no limit to the amount of work one can put into a lawn.

If you demand a weed-free lawn resembling a putting green, you may have to spend some time there nearly every day and long hours on week-ends mowing and pulling weeds and keeping insects under control.

Plan Maintenance Carefully

In the first stages of garden planning the homeowner should try to make a decision about how much time he is willing to spend on a lawn. There are several kinds of lawn grasses and they vary in the amount of care they demand. The Bermuda grasses are the most demanding, requiring, as turf experts say, "a high level of maintenance." Zoysia comes next, requiring a moderate amount of maintenance, while St. Augustine is usually put into third position. Centipede is fourth down the list, not only requiring but demanding a low level of maintenance. The Bahia and the mixed seed lawn grasses are also looked upon as low maintenance grasses.

Despite its susceptibility to chinch bug injury, St. Augustine is by far the most popular lawn grass in south Florida. Many discouraged homeowners who switched to other grasses after St. Augustine was destroyed by chinch bugs later returned to St. Augustine. For this grass tops all others in its tolerance of neglect and irregular mowing, resistance to cold and drought, and tolerance to shade. Several varieties of St. Augustine are planted, the most popular of which in recent years has been Bitter Blue. A new variety, Floratine, developed by the University of Florida, promises to be superior to any other variety. It is a vigorous, short-jointed grass which hugs the ground and which may be mowed somewhat shorter than Bitter Blue St. Augustine.

The chinch bug is the only major problem anyone is likely to have with St. Augustine, but control is fairly easy if an insecticide is applied when the first signs of these pests are noted. The presence of chinch bugs should be suspected when any yellow area is seen in the lawn. The yellowing areas grow larger if neglected, and soon much of the lawn may be dead. If the homeowner intends to do his own spraying for chinch bug control, he should learn to recognize this small black insect, about one-fifth of an inch long, with white cross markings on the wings. Chinch bug colonies increase and become more widespread in their feeding. A number of insecticides are now available which give very good control and which may be applied successfully by the homeowner. Directions must be carefully followed if the insecticides are to be effective. They have been prepared from information compiled by experiment stations and through the work of entomologists and commercial spray operators over a period of several years. One of the most important directions every manufacturer's label carries advises the homeowner to be sure that the lawn has been thoroughly watered before the insecticide is applied. It is best to run sprinklers until about an inch has been put down. Then, after the insecticide is applied, the sprinkler system is cut on again for a few minutes to wash the chemical through the thick mat of grass.

Probably most widely planted next to St. Augustine are the several varieties of zoysia, of which Emerald zoysia and Meyer zoysia (Z-52) are the most popular. Emerald is somewhat finer and is said to be more vigorous. But when they are planted side by side it is difficult to tell much difference between them, except in the texture. Both varieties must be propagated from sprigs, plugs, or full sod. Zoysia is a slow growing grass, and if planted from sprigs, as it is often, the sprigs should be set as close together as one has money, time, and patience to do. Five or six inches apart is not too close.

Growth is speeded if the soil is moderately rich, and if monthly applications of fertilizer are made. It is very important to keep zoysia free of weeds, and especially of wild Bermuda, while it is getting established. During the growing season it must be mowed regularly, usually once a week. Only a reel type mower is recommended for this rather tough-bladed grass.

Centipede Needs Little Care

Centipede, often called "poor man's grass," has its ups and downs in popularity. The main need for this grass is irrigation during dry periods. If the soil has any richness at all, centipede may not need to be fertilized. One application should be made in the fall, at half the normal rate recommended for St. Augustine and other grasses. In very poor soil a second application may be made in the spring. This should not be done, however, until it is evident that neither rainfall nor irrigation alone will bring out the green color in the grass. Centipede does not live in shade nor will it tolerate salt spray. It thrives best on acid soil, but will tolerate limestone soil. An application of an iron compound can be made if grass becomes yellow. This should not come in contact with walks, walls or patios, as staining will result. Either iron sulphate or neutral iron may be used.

A mixed seed type of lawn grass has become popular in recent years. The mixture usually contains Pensacola Bahia, centipede, and

Zoysia japonica. Is also contains seeds of quick-germinating northern type grasses which may thrive through winter, but which die out after a few weeks if planted in summer. These mixed grasses seem to be satisfactory for persons who do not demand a fine turf and who do not find the mixtures of textures objectionable.

The improved Bermuda grass varieties make the finest turf of any other kind of grass planted in Florida. Bermuda is not for the week-end gardener. One must really want a fine turf in order to do a good job maintaining a Bermuda lawn. The fastest growing of all locally planted lawn grasses, Bermuda requires the richest conditions, and it also requires mowing more often than any other grass. Only the select varieties should be planted vegetatively from sprigs or cuttings. Probably the most widely planted Bermuda is T-57, sold also under the name Tiflawn. It is a durable grass and is widely used for racing courses, baseball and football fields, and for golf fairways and tees. Tiflawn is too coarse for golf greens. Another select variety, T-127, or Tifine, is used for putting greens. Also available for planting in south Florida are T-328 or Tifgreen, and T-419, or Tifway. They are all excellent, but they do well only when they receive the care any Bermuda lawn should get. African Bermuda variety, Uganda, also is sold in south Florida, but it is patented and is offered only by nurserymen licensed by the patent owner.

Expert Attention Needed

Anyone wishing to develop a Bermuda lawn must either learn a great deal about lawn development and maintenance, or must have the services of an expert to prepare the soil, to do the planting and the maintaining of the grass afterwards. Bermuda requires constant attention. You cannot plant it and then go away for the summer and expect to have a fine Bermuda lawn when you return — unless you have made arrangements with a reliable maintenance man, to mow, fertilize, and keep insects under control. For during the growing season Bermuda may require mowing twice a week, preferably with a precision mower. It may require up to four applications of fertilizer a year. It must be watered weekly during long periods of drought. If sod webworms are not controlled these pests can be as bothersome in a fancy Bermuda lawn as chinch bugs are in a St. Augustine lawn.

Bermuda does not thrive in shade and should not be used if there is a great deal of deep shade in the garden. On the other hand, where there is enough light for zoysia, Emerald zoysia can be planted in the shade and the Bermuda planted in the sun. If mowed regularly very little difference between the two grasses will be noticed. Where shade is so deep that zoysia will not grow, ground covers, such as peperomia or mondo grass, may be planted. Or, the owner may prefer to let leaves lie where they fall, making a brown blanket over the ground. The more orderly-minded person may want to pour a dividing strip of concrete in a circle around trees to keep the mulched areas separated from the green turf.

Be wary of planting Bermuda where there is a lot of nut grass, unless the soil can be fumigated and the nut grass killed. Nothing else can be depended upon to do a better job of spoiling a fine Bermuda lawn. The aggressive spikes of the nut grass shove through the lawn immediately after it is mowed, making themselves so objectionable that they are seen before one sees the lawn itself.

Before any kind of grass is planted the soil should be well prepared. All pieces of mortar, timber, tile, and concrete blocks should be removed before soil is put down. Only moderately rich soil is required for most lawn grasses. It can be improved through use of fertilizer, which provides nourishment at the same time. Muck should be avoided. It is often full of nut grass and other weed seeds; and muck is not long lasting anyway, as it quickly oxidizes. Large quantities of marl may cause nutrition problems in shrubs and trees later. Perhaps a small amount of marl mixed well with the soil would do no harm. A dark sandy loam should be preferred, but this type of soil is usually very hard to get. Pit sand is cheaper, and it can be improved sufficiently by digging in a couple of inches of pulverized peat moss, along with fertilizer spread at the rate of about forty pounds per thousand square feet. On old lawns, that are to be replanted, two inches of pit sand can be spread over the area, after the weeds have been sprayed with a weed killer, and planting can be done in the sand after it has been fertilized and wet down well.

Grasses may be planted by sprigs, plugs, full sod, or, in some instances, by seeds. All improved varieties must be planted vegetatively. They do not come true from seeds. Sprigs are recommended over plugs. It is best to have grass sprigs lifted with the aid of a fork rather than obtained by tearing up pieces of cut sod. Many persons think that they can solve their grass planting problems by putting in full sod. But the solid planting of sod has its problems too. It requires several weeks for the roots of the thin sod to penetrate into the soil and become established. The newly planted sod cannot be neglected during this time, or the grass will suffer. And if a lawn is planted with sod coming from mucky areas, additional troubles may be in store. Muck-grown sod appears to take longer to get established, and in the meantime the muck itself oxidizes and disappears, leaving a very uneven surface which must be leveled with the aid of repeated applications of sand.

Propagation Methods

Propagation by sprigging has many advantages, and is the method generally recommended for Bermuda and zoysia. Centipede and St. Augustine are more often planted from squares which are cut from pieces of sod. In rocky soil, however, it is often difficult to dig holes for the planting of squares or plugs, and so it might be much easier to tear the pieces of sod apart, shake the soil from them, and plant individual sprigs. This is easily done. The sharp point of a pick is used to make holes, as close together as is desired, and the sprigs are placed in these holes. If a sprig of grass eight inches long is to be planted, then at least half of it should be stuck into the hole made by the aid of the pick. Do not press the soil about the sprigs with the aid of the fingers. Instead, use a garden hose and wash the soil into the holes. This will firm the soil about the sprigs and avoid the creation of air pockets. The sprig becomes established quickly with a minimum amount of watering. Seldom is a top dressing of soil or sand necessary later to even the surface.

If seeds are to be used, the soil should be well prepared before hand, and all debris removed from the carefully leveled surface. If the soil is rocky an inch or two of sand can be put down for the seed bed. This should be fertilized heavily a few days before planting, and wet down well so the fertilizer is dissolved. If the soil beneath the sand is exceptionally poor, an inch or so of pulverized peat moss may be put down at the time the fertilizer is applied and both raked into the sand. Plant seeds according to directions on the package and carefully rake into the top surface of the sand. Watering is a must. If the seeds dry out they will die. Watering must be continued frequently until the young grass plants are well established. If directions on the seedman's package are followed no problem should arise.

How to Apply Fertilizer

Grass varieties vary in the amount and number of applications of fertilizer they demand. Centipede requires very little fertilizer while Bermuda may require a great deal if it is to be kept green throughout the year. Many homeowners over-fertilize. Usually St. Augustine requires but two applications a year, and probably should receive only one application. If one application is made it should be done in the fall, after the end of the rainy season, about the first of November. If necessary a second application can be made about the end of March. Grass should never be fertilized when the soil is dry. If grass must be fertilized at a time when there is no rainfall for some time, it should be well irrigated the day before. After the fertilizer is applied the grass should be irrigated lightly, to wash the fertilizer off the grass blades and into the soil. If fertilizer is applied when the soil is excessively dry, and when the grass may be in a stage of near wilt, it will take up large quantities of fertilizer salts after the fertilizer is watered in, and injury may result. When watered the day before, the grass will be able to take up only a small amount of salts, as the tissues will already be filled with water. Thus possible injury is prevented.

So many kinds and mixtures of fertilizers are on the market today that it is hardly possible to give an over-all recommendation of how much to apply. Manufacturers are now including directions on the fertilizer bag, or on the label, and these directions should be carefully followed. Generally, a 6-6-6 analysis fertilizer is applied to grass at the rate of twenty to twenty-five pounds per thousand square feet. A light application would be half this amount. The gardener will save money in the long run by buying a fertilizer containing a high percentage of organic nitrogen. Organic nitrogen does not leach out of the upper levels of the soil during heavy rainfall or irrigation, as do the quickly dissolved nitrogens, which may be somewhat less expensive than the organic type.

Fertilizing in the summer should be avoided whenever possible, except for Bermuda which may need one or more applications between June and September. For other grasses the spring application should carry over very well, especially if the nitrogen used is of an organic type. One way to tell whether grass needs fertilizer is to water well for a couple of weeks, if watering has been neglected, and see whether the grass develops a rich green color. If the grass doesn't turn as green as neighboring lawns, many homeowners want to fertilize right away. This can be a mistake. But everyone who follows this practice of trying to keep up with the Jones' green lawn, should be forewarned that growth will be rapid, mowing must be done more often, and insecticides must be kept ready for sod webworms and armyworms which are sure to take advantage of the lush, rich pasture so generously developed for them. For it is a fact that the more you fertilize your lawn the more work your lawn will require. If mowed irregularly, you are liable to scalp the fast growing grass. You must top-dress with sand once a year at least to take care of the dense mass that centipede and St. Augustine develop when grown under rich conditions.

Watering Practices Important

Watering of lawns is a subject on which there are many opinions, and many different practices are followed, from watering a little every day to watering not at all. One extreme sometimes can be as bad as the other. The turf experts agree that watering once a week during dry periods usually is enough. Probably the ideal system of watering is to start the sprinklers after the first sign of wilt. At least one inch of water is put down, and no more watering is done until the first sign of wilt is noted again. Grass when wilting will turn slightly gray in color. To determine how long to run sprinklers in order to put down an inch of water is very simple. Place several shallow pans about beneath the sprinklers and see how long it takes for an inch to collect in the pans. This test also tells one whether there is even distribution of water. Daily light sprinkling is objectionable because when only the top inch or two of the soil is moistened, most of the grass roots develop near the surface. Periodic deep watering encourages deep rooting. Grass with deep and extensive roots obviously can stand neglect, as when one is called away on a trip. Shallow rooted grass declines quickly when neglected.

Lawn grass is seldom mowed frequently enough although no time table can be set. It must be mowed when mowing is needed. At one time of the year it may be weekly while at another time of the year once a month may be enough. The best advice to give is to mow frequently enough to keep the blades growing more or less parallel with the ground. If mowing is neglected during the lush growing season, and especially after grass has been fertilized, the crowded grass blades leave their horizontal position and turn upward, seeking the sun. Unless the mower blades are set to cut higher than normally, it is obvious that most of the grass blades will be cut off, with possible injury to the lawn resulting. In the case of centipede and St. Augustine, green stems may be exposed, and if the mowing is done during a dry period the exposed stems may burn and die beneath a scalding sun. Both Bermuda and zoysia may turn brown after close mowing following a period of neglect, but they are likely to recover because these grasses have underground stems, while centipede and St. Augustine, whose stems are above the ground, may not recover. During periods of drought, grass should always be irrigated the day before it is mowed.

It is impossible to give exact information on just how high any grass should be cut. The height of cutting depends on how often the grass is cut. Regularly cut grass can be cut lower than the grass which is cut at rare intervals. Golf putting greens are mowed daily. An expensive precision mower is used, and is adjusted usually to cut about one-fourth of an inch high. But what is good for putting greens may not be good for lawns. Bermuda lawn grasses are cut at a height of about three-fourths of an inch, but if mowed frequently enough could be kept at one-half inch. Zoysia is cut from three-fourths to one inch, centipede at about one and one-half inches. St. Augustine may be cut somewhat higher. A good rule to observe is to always set the mower so that no more than one half of the grass blades are cut off.

Which Lawnmower?

The homeowner should buy as good a mower as he can afford. For most grasses a reel type mower will be found superior to a rotary type. The rotary is all right so long as the blade is kept sharp, but usually the blade is seldom sharpened, and it becomes so dull that grass is frayed rather than cut. Moreover, a rotary mower is a very dangerous instrument and should be used with extreme care. It is advisable to cut off the motor and remove the spark plug connection before getting the fingers close to any mower blade, whether rotary or reel. And many a finger has been lost or severely mutilated where the operator got his hand too close to the blade of a rotary mower while trying to remove a piece of trash tangled in the blade. The motor should be cut off before any attempt is made to untangle weeds or debris from a revolving blade.

There is no sure way to prevent weeds from getting started in lawns. It is best to start out with a lawn free of weeds and weed seeds if possible. Weeds, and weed seeds, especially nut grass, are often introduced in top soil. If muck is to be used – and generally it should not be used – it should be fumigated to get rid of the weeds. But no matter what precautions are taken, weeds will get into any lawn, waiting to take over as soon as the homeowner turns his back.

Weeds have a tough job taking over a well kept lawn. If kept in fair condition, lawn grasses will crowd out most weeds. And if weeds threaten to take over in spots they can be removed before they become too well established.

Chemical weed killers should be used with a great deal of caution, and the user should understand full well the dangers of using them. Directions for use must be carefully followed. The hormone weed killers, 2,4-D and 2,4,T, are safe enough to use on fine bladed grasses such as Bermuda and zoysia. They must be used with more care on the wide bladed grasses, especially centipede and St. Augustine. St. Augustine lawns often may be severely injured through the use of weed killers, and for this reason many manufacturers will not recommend 2,4-D on this grass. Many persons do use the hormone weed killers on St. Augustine grass and report good results, so it obviously can be used safely. If directions on the label are carefully followed perhaps little harm will be suffered.

Never apply weed killers when the soil is excessively dry. Don't expect weeds to die immediately after 2,4-D is used. It may be several days before any change is noted. Weed leaves may simply turn yellow and decline, or they may become distorted and decline. Sometimes several applications are necessary to kill weeds. The hormone weed killers will depress nut grass in Bermuda and zoysia grass but may not kill it out completely, except after repeated applications of a year or more. Such treatment to centipede and St. Augustine would not be possible without also killing the lawn grass. Most weeds may be kept under fair control through regular mowing. Wherever weeds begin to form patches and take over areas of grass they should be quickly removed. Usually satisfactory weed control is possible through this method alone.

Sand, or top-dressing of soil, is recommended for one purpose – to level uneven lawns and to overcome the spongy effect which develops in centipede and St. Augustine lawns. A native pit sand or a mineral type sand used by builders is usually recommended, because high grade, weedless top soil is very hard to find. Neither muck nor marl is recommended, nor is the very fine limestone which is a by-product of crushing gravel.

Control Lawn Pests

Insects are usually the homeowners' greatest worry in lawn maintenance. No grass is immune to pests. In recent years the chinch bug has been a major hazard in the growing of St. Augustine grass, but much of the trouble has been brought on by homeowners themselves, who have tried to out-do one another in seeing who could have the greenest lawn. Chinch bugs are especially hard to control in a lawn where there is a thick sponginess. They are much easier to control in grass grown in moderately rich soil which is moderately fertilized. Removal of grass clippings also helps to prevent thatching, which creates the deep sponginess one notes in so many St. Augustine lawns.

Several chemicals are on the market which will give good control of chinch bugs, provided the manufacturer's directions are carefully followed. Control of sod webworms and army worms likewise is very simple. Apply the insecticide after the first sign of insect damage. On centipede and St. Augustine the damage will make itself known by the disappearance of the grass blades. The lawn will appear to have been closely mowed in areas where the insects have chewed. With Bermuda and zoysia the damage may appear as brown spots in the early stage of feeding, with the spots becoming larger daily. For this reason it is necessary to have insecticides on hand so that they may be applied immediately after the first sign is noted. If these grasses are fertilized during the warm season, when these insects are more likely to be active, it is a good idea to apply an insecticide about three days after the fertilizer has been applied, as a preventive measure.

Several diseases attack grass, but for the homeowner to be able to identify them would require long study and close observation. One of the most common of fungus diseases causes yellowing and then browning of grass. If the blades of grass are carefully observed along the borders of declining areas, small yellow spots or streaks will be noted, where the initial infection begins. These areas spread throughout the leaf until it turns yellow and then dies. Another disease may cause large patches of the lawn to appear as though someone had poured scalding water over the grass. At other times, and especially during long periods of humid weather, a gray mold may even develop in large areas of a lawn. Several fungicides are now available which give good control of grass diseases. A fungicide, or a combination of fungicides, should be applied immediately after the first sign of disease is noted. Daily watering, especially in the evening, may encourage the development of grass diseases.

It should be remembered that the more fuss and bother a lawn receives the more troubles are likely to develop. Fewer troubles are likely to develop in a moderately maintained lawn than in either a highly maintained or a neglected lawn.

Wedelia is widely used ground cover for ocean front gardens

Gravel may substitute as "ground cover" in special areas

Chapter 12

GROUND COVERS

The use of ground covers has become so important in garden design in recent years that one could hardly visualize a contemporary garden without them.

Ground covers, of course, are not new. Grass is a ground cover. So is ivy, widely used in the north. The broad-leaf ground covers used to be reserved for shady areas where grass was difficult to grow, or for rough hillsides where the cultivation of other plants was not practical. But today we use ground covers under all kinds of conditions. Landscape architects use them to create a new kind of form in the garden, in complicated designs. Several kinds of colors of ground covers may be joined through the use of free-flowing lines, creating mosaic patterns that would have been unthinkable in the formal garden of a few decades back.

The use of masses of ground covers in open areas is believed to have been started by a Brazilian artist and landscape designer, Roberto Burle Marx. A master plantsman as well as a master designer, Marx hit upon the idea of using the adaptable native materials around him which he could propagate in large amounts. His design spread to Europe and America, and is especially popular in Florida and California.

The complicated design used by the contemporary landscape architects is not easy to attain. The result is either good or bad. There is never any half-way. The inexperienced garden planner would do better justice to his own place by sticking with the simplest lines, and avoiding any attempt to make mosaic patterns with several kinds and colors of ground covers confined within fancy curlicues.

What Makes a Ground Cover?

Almost any low growing plant which produces a mass of foliage is suitable for ground cover planting, but some plants are easier to grow and some require much less care than others. The use of those which are difficult to grow usually should be avoided. Plants which normally require the protection of a greenhouse, such as the calathea, episcia, fittonia and unusual kinds of peperomia should not be thought of. There are scores of more suitable plants.

Probably the most widely used ground cover in South Florida is wedelia. It is easy to propagate, it grows rapidly, it quickly shades out weeds and smothers them, and it is easy to maintain. Wedelia will survive light frost with little injury, and even a heavy frost may injure only the top part of the plant. Other popular ground covers are zebrina, blood leaf (*Iresene lindenii*), rhoeo, purple queen variety of setcreasea, mondo grass, liriope, lantana, various ferns, dichondra, kalanchoes, bromeliads, aloes, ajuga, malpighia, the sprawling varieties of carissa, as well as such bulb plants as crinums, spider lilies, amaryllis, and daylilies.

In shady locations various kinds of aroids may be used, including pothos, syngoniums, and the vining kinds of philodendrons. These plants will attempt to climb trees but, can be discouraged by pulling the climbers off the tree trunks every couple of months. The common asparagus ferns, *Asparagus plumosus* and *Asparagus sprengeri,* might well be used for ground cover planting. *Asparagus sprengeri* may be grown in immense masses about larger plants such as self-heading philodendrons, with which the foliage color and texture make an interesting contrast.

Even certain trees and large vines may be trained as ground covers for special situations. The sea grape is an example of a suitable ground cover where one wishes to have a low mass of large foliage. By selecting and airlayering short, sprawling types of plants, and planting them close, it is possible to create a mass effect in a single bed. No plant should be more than a foot high after planting. It is necessary only a few times a year to wade through the mass or use a long handled tree pruner, to cut off new branches which tend to grow vertically. Such plantings are especially suitable for sea shores and other places where salt tolerant plants are required.

There is no reason why the common varieties of bougainvillea could not be used for ground cover planting. Small, low plants should be used, and these may be trimmed back each spring or early summer, after the winter flowering season is over.

The kinds of ground covers one plants may be determined by a variety of things, including the preferences of the gardener, the kind of soil, the amount of sun and shade, the type of garden plan, and plant availability. Individual plants may cost fifty cents to a dollar each. Thus if one intends to plant large areas of certain ground covers, the cost can quickly run pretty high. Plants which may be grown from cuttings, such as wedelia, or which are grown from division, such as the mondo, liriope, and ferns, are the most widely used. Even so, a hundred plants of mondo or liriope do not go very far. But one can start small areas and wait for the plants to grow and increase.

Ground cover plantings are seldom static things, remaining indefinitely the way they are planted. One tires of wedelia or zebrina and wishes to replace them with something else. Even the rich colors of purple queen and blood leaf eventually become too much for the most rabid, color-hungry gardener. Thus it is always good to be looking ahead, and collecting and growing different kinds of ground covers in order to have material for replacing those of which we tire. Very often we can trade with friends when they thin out or renovate. There is, however, one danger in exchanging ground cover plants with friends – the possibility of introducing nematodes.

Boulders Now Popular

The use of large boulders in garden design in recent years has become so general that we seldom see a contemporary garden without a display of huge rocks. They are almost always associated with ground covers. The boulders are set so that they appear to be outcroppings, as one finds them on hill sides in mountainous country.

It is not difficult to see the Japanese influence here, but no attempt is made to carry Japanese symbolism with the design. In Japanese gardens, large boulders outcropping from sand may represent mountains rising from the sea. Undulations are carefully raked into the sand so that little imagination is needed to picture waves pounding against rocky shores. If the American landscape designer were asked why he uses boulders, his only answer could be that he uses them because they are aesthetically pleasing. When used correctly, they are. What is correct and what is not correct is often difficult to put your finger on; but the attempt by amateurs to mix boulders and ground covers usually ends up in what might appear to be a rock pile dumped by mistake in one's garden.

Boulders may be used in a number of situations, in dry gardens where desert scenes are represented; in shady locations where ferns and mosses may grow on the surface of boulders subjected to a constant trickle of water; or in large mosaic beds of many colorful ground covers, laid out in free-flowing lines.

There are ground covers for every situation, and we are especially fortunate in having a number which thrive out in the full, glaring sun under conditions intolerable for so many plants. Outstanding among these are the aloes and the kalanchoes. These are especially suitable for growing in dry bed locations with agaves, yuccas, and euphorbias. Such planting often give a cheerful note to the subdivision home where otherwise you see only small switches growing for block after block.

But these are not the only kinds of ground covers which might be used to brighten up the sterile lot about a subdivision home. Wedelia, purple queen, rhoeo, crown-of-thorns, and the periwinkle all thrive in full sun, if given a little help by providing them with suitable growing conditions.

Thus the ground covers stand out foremost of all the ornamentals in their ability to provide life and color in the new garden, and quickly. It is true that in time one may eventually tire of the gay colors and in the plant forms themselves, but in the meantime the permanent garden's shrubs and trees should be coming along. In time the old ground covers will give way to new situations, new forms, and new ideas.

The many kinds of ground covers grow in a wide variety of soils. Most of them do require a fairly well drained situation and are likely to suffer severely if they must be subjected to long periods of flooding or soggy soil. The kalanchoes and aloes should be planted in a sandy mixture, in a bed which is elevated somewhat above the surrounding ground level. Rather course sand is recommended. To provide a feeling of desert-like conditions, two layers of sand may be put down. In the first layer is incorporated peat moss or good soil, with a small amount of sheep manure added. The humus need not be very much, constituting only ten to fifteen per cent of the sand. On top of this another layer of an inch or two of sand is placed, but no humus is incorporated with this. The plants are then set out so that the longer roots come in contact with the lower soil level. Or, if the plants are not sufficiently well rooted for planting them so deeply, the roots will quickly reach to the lower level of their own accord.

Such ground covers as wedelia, purple queen, and zebrina require fairly rich soil, but it should be well aerated. The practice that one occasionlly sees of using only peat moss as a mixture for planting ground covers leads to trouble. This material remains soggy for long periods after irrigation or rainfall, and roots suffocate. You can usually provide an excellent soil situation by spreading a couple of inches of peat moss over the surface, plus about four pounds per one hundred square feet of a balanced fertilizer such as 6-6-6. If the soil is sandy the humus should be mixed with it to a depth of at least six inches. Mixing humus with limestone is hardly possible. A pulverized composted material should be spread over the surface before planting. But under such conditions fertilizer must be applied several days before planting time and washed down into the soil with a thorough irrigation. The soil should be disturbed only through the mechanics of planting.

In general, the growing of high quality ground cover plants is no different from growing anything else. Plants need to be fertilized occasionally and watered when they need it. They should be kept relatively free of insects and diseases. Slugs are especially serious among many ground covers. These are easily controlled with the help of metaldehyde slug bait. Sometimes diseases, especially during rainy weather, will get started in dense ground covers. A fungicide or antibiotic spray may be used if the trouble becomes severe.

In the list of ground covers below, the performance of each plant in sun or shade is noted, as well as recommended methods of propagation. Where possible, it is always a good idea to get a few roots on cuttings before placing them in permanent beds. The cuttings can be set very close together in flats. These can be kept in a semi-shady location until rooting is assured, and then they can be gradually placed in the sun to harden up before being planted in a permanent location in unprotected sunny beds.

The list:

AJUGA. Bugle weed. A bronze-leaved plant reminding one of leaf lettuce in form, reaching six inches in height. Sold locally under the name of *Ajuga reptans.* Althought ajuga is a northern plant, it thrives in south Florida under good care. Must be grown in nematode-free soil, and does better in part shade than in full sun. Propagated by division.

ALOE. Numerous species of this interesting plant from the dry areas of Africa are in cultivation, and additional ones are being introduced. It has a definite place as a ground cover plant, both because of its unusual fleshy form and because of its colorful bloom which persists for several weeks in late winter and spring. Seeds or off-sets. (See Annuals and Perennials.)

ASPARAGUS SPRENGERI. Asparagus fern. This plant widely used for making hanging baskets, also may be used for ground cover planting. Thrives in full sun or part shade, but does its best in moist, fairly rich soil. A relative, *Asparagus plumosus*, also may be used as a ground cover. Pruning would be required occasionally to keep this species from climbing. Division.

ASYSTASIA COROMANDELIANA. Coromandel. This viney plant makes a dense ground cover but it also is an aggresive climber. The tubular, light purplish flowers appear several times a year. Grows in full sun or part shade. Tends to become a weed. Cuttings.

BROMELIADS. Many of these plants make desirable ground covers under some conditions, but most of them require at least part shade. (See *Annuals and Perennials.*)

CALADIUM BICOLOR. A very colorful plant, but beds must be remade from time to time as tubers become dormant and tops die. Can be grown in full sun in winter, but during the summer leaves tend to burn unless the soil is kept quite moist. Tubers.

CRINUM. Several species are grown in Florida, but most of these plants are too large for use as ground covers on a small place. They are instead better used as substitutes for low shrubs. *Crinum asiaticum* is probably the most widely grown in Florida. Division.

CUPHEA HYSSOPIFOLIA. This is a low green shrub with very small leaves and producing large quantities of light violet flowers throughout the year. A very attractive ground cover, but because of its susceptibility to mite injury and its intolerance of drought, has lost much of its popularity. Seeds or cuttings.

DICHONDRA CAROLINENSIS. A creeping herb often used as a substitute for lawn grass. In south Florida it does somewhat better in part shade than in full sun, except in deep soil which is kept fairly moist. One objection to dichondra is its susceptibility to fungus, which is capable of killing large areas of plantings in a few days. It is one of the few herbacious plants which can withstand tramping like grass. Seeds, cuttings, or division of sod.

DOLICHOS HOSEI. Sarawak bean. This small but aggressive viney bean-like plant makes an excellent ground cover under some conditions. Its main objection is its persistent invasion of lawn grass, much the same as dichondra. It is capable of surviving close and frequent mowing of the grass. Where permitted to grow freely it makes a dense mat six to ten inches deep and, being a legume, no doubt is an efficient soil builder. It is susceptible to mites, caterpillars, and to cold damage. The Sarawak bean is not recommended for use as a ground cover except possibly for groves. It does not thrive in deep shade. Cuttings.

EUPHORBIA MILII. *Euphorbia splendens*. Crown-of-thorns. This well known colorful flowering plant, armed along its branches with sharp spines, deserves wider use as a ground cover. It makes a formidable border. Its bright red flowers are produced almost continuously throughout the year. Cuttings.

FERNS. Only one fern, *Nephrolepsis exaltata,* is widely used as a ground cover in south Florida. This native Florida fern is better known as Boston fern, of which there are several varieties. It will grow in full sun, but does best in shade. Once established it will grow under the most neglected conditions. Division.

FICUS QUERCIFOLIA. Oak-leaved fig. A fairly new plant to Florida horticulture, with large leaves resembling those of a white oak. It is a fairly rapid grower and should find good use as a ground cover. Thrives in part shade. Cuttings.

HEMEROCALLIS. Daylily. This is an excellent flowering plant for south Florida gardens and deserves wider use, both as a border plant and as a ground cover. It thrives best in full sun, although plants will grow and flower fairly well in part shade. Flowers persists for several weeks in late spring and early summer. Several varieties are available. Division.

HYMENOCALLIS. Spiderlily. Although this bulb plant is somewhat large for use as a ground cover in small gardens, it might be used under some conditions. The native Florida species, *Hymenocallis keyensis,* may be planted among the front line salt resistant plants along the beach. Division.

IRESINE LINDENII. Blood leaf. A highly colorful herbaceous plant whose common name indicates its hue. Demands full sun for best effects. Too rich in color for most gardens, althought it is widely used. Cuttings.

JUNIPERUS CONFERTA. Shore juniper. The only conifer which is used as a ground cover in Florida. It is one of our more interesting plants, and is often used about boulders, where its long streamers are permitted to spread over the surface of the rock. Cuttings.

KALANCHOE. Bryophyllum. Several species of kalanchoe are in cultivation in Florida, and a few of them are widely popular ground covers. Probably the most popular is the *Kalanchoe fedtschenkoi,* and probably next to it is *Kalanchoe gastonis-bonneri. Kalanchoe tubiflora* and *Kalanchoe longifolia* are escapes from cultivation in south Florida and are commonly seen growing over vacant lots. Leaf cuttings.

LANTANA CAMARA. A sprawling shrub with yellow or orange flowers turning red or scarlet. Dwarf varieties are commonly used for ground covers in Florida. Grows well in pinewoods. Cuttings.

LANTANA MONTEVIDENSIS. *Lantana sellowiana.* Trailing lantana. A common shrub in south Florida somewhat larger than *Lantana camara,* and producing large numbers of rose-lilac flowers. Often seen growing wild. Cuttings.

LIRIOPE SPICATA. Creeping lilyturf. A grass-like plant, but member of the lily family, widely used throughout the southeast as a ground cover. It is an excellent shade plant, reaching about a foot in height, and quickly makes a dense cover if plants are set about six inches apart. It is highly resistant to drought and is capable of withstanding long neglect. Division.

MALPIGHIA COCCIGERA. Holly-leaf malpighia. Several sprawling varieties of this low shrub have become very popular for ground cover planting. Small pink flowers are borne profusely throughout the warmer months, and these are followed by small red fruits. Cuttings.

OPHIOPOGON JAPONICUS. Mondo. Mondo grass. Dwarf lilyturf. This dark green, grass-like plant resembles lirope but is much smaller. Like lirope, it should be grown in shade, as it suffers top-burn during the summer if planted in full sun in south Florida. Sold under the name of "mondo grass," this plant is often represented as a substitute for turf grass, which it is not. It does not stand tramping. Mondo, however, is one of our most outstanding ground covers when used in shade. Division.

PEPEROMIA CRASSIFOLIA. This dark green, fleshy plant is an excellent ground cover for shady locations. Cuttings.

PHILODENDRON. Several of the vining types of the philodendrons make excellent ground cover plants when grown in shade. Because they tend to climb tree trunks they are sometimes looked upon as troublesome. However, there should be little work in pulling the runners off tree trunks two or three time a year.

PHYLA NODIFLORA. Lippia. Matchweed. This viney plant, mistakenly called "creeping Charlie" in south Florida, is looked upon as a pest in lawns where it commonly occurs. It makes an excellent ground cover, though, and, like grass will withstand tramping. Occasionally one sees an excellent lawn of lippia, where chinch bugs have previously killed the St. Augustine grass and this "weed'" has been permitted to take over through neglect. Cuttings.

PILEA CADIEREI. Aluminum plant. A low herbaceous plant with broad bands of bright aluminum-like color on its green foliage. Thrives best in part shade. Cuttings.

PILEA MICROPHYLLA. Artillery plant. This is often a common weed in gardens where, once started, it is never eradicated. Seeds abundantly.

PLUMBAGO CAPENSIS. This low-sprawling shrub deserves wider use as a ground cover. Blue or white flowers. (See *Shrubs and Their Uses.*) Cuttings.

POTHOS. See *Scindapsus aureus.*

PYROSTEGIA IGNEA. Flame vine. Although this aggressive vine with its crimson-orange flowers is often seen on the roofs of Florida homes, it grows very well as a ground cover where there is lots of room in full sun. It does not thrive in shade. Runners can be pulled down periodically from trees, shrubs, or fences. Cuttings.

RHOEO SPATHACEA. *Rhoeo discolor.* Oyster plant. Moses-in-the-cradle. This herbaceous plant, with broad sword-like leaves, deep green above and purple beneath, is among south Florida's most adaptable plants. It does well in all kinds of soils and situations, in shade or in full sun. It deserves wider use. Division, or seedlings which volunteer in large numbers near older plants.

SCINDAPSUS AUREUS. Pothos. This large-leaved vine makes a dense ground cover where it can be grown in shade. Two varieties, one with deep green leaves and another with green leaves splotched with yellow, are in cultivation. Vine cuttings.

SETCREASEA PURPUREA. Purple queen. A deep purple, herbaceous plant which has become one of south Florida's most popular ground covers. It is oftern over-used and poorly used, although no one can deny that immense beds of this plant are sometimes unforgetable, a fact which may be unfortunate. It is a good plant but should be used with restraint. Cuttings.

SYNGONIUM. This aroid, like the philodendron and the pothos, makes an excellent ground cover in shady areas. Vine cuttings.

TRIBULUS CISTOIDES. Caltrops. Bur-nut. A sprawling herbaceous plant which clings tightly to the ground, but when grown in dense numbers tend to make a mass several inches deep. Its yellow, buttercup-like flowers are produced throughout the warm months. The main objection to this plant is its very spiney fruit, which is capable of puncturing a bicycle tire or the sole of a shoe. Not recommended where there are children. Seeds or cuttings.

VINCA ROSEA. Periwinkle. A herbaceous shrub which deserves wider planting as a ground cover. Two varieties, one with rose-purple and one with white flowers, are common. Seeds.

WEDELIA TRILOBATA. A rapid growing herbaceous plant, bearing large numbers of yellow, daisy-like flowers throughout the warm months. It is one of south Florida's best ground covers. Recommended for full sun. It will thrive under brackish conditions. Cuttings.

Plants are living things. Each has a personality. And, although plants have no nervous system, one wonders if a majestic tree doesn't feel some indignity at having to associate with scrawny, low-class plants many gardeners place beneath their protective branches, like buzzard biddies huddled beneath the wings of a graceful swan.

Impatiens is a bright flowering annual in tropical gardens

Tropical perennials take the place of northern perennials

Chapter 13

ANNUALS AND PERENNIALS

Many of the annual flowers which thrive in northern gardens can be grown in south Florida during the cool part of the year, but perennial flowering plants, with few exceptions, must be confined to the tropical species. How well one grows any plant in south Florida, whether it is annual, tropical perennial, hardwood tree, or shrub, depends to a large extent upon one's ability as a gardener. But the successful growing of annuals also depends upon the time of year that planting is done.

In order to grow outstanding annual flowers, the gardener must recognize two things: first, that planting should be done at the beginning of the cool, dry season – after about the middle of October – rather than in the spring as in the north. While it is true that a few may be grown in south Florida throughout the year, insects and diseases become such hazards during the late spring and summer that one must engage in a constant battle against these pests. Furthermore, if plants come into bloom during a rainy period they are likely to be very poor and flowering lasts but a few days. Secondly, the gardener must recognize that south Florida soils are extremely poor and that if they are not enriched the planting of annuals is hardly worthwhile. Humus, such as peat moss, along with a balanced fertilizer and pulverized sheep or cow manure, must be incorporated with the soil. When available, a mixture of muck and marl may be used to make the annual flower beds, but the muck disappears after a year, due to oxidation, and the humus must be replenished, preferably by digging in a couple of inches of peat moss. Soil which is free of pests and weed seeds should be used whenever available.

Veteran south Florida gardeners generally do not go in for annuals. Annual growing is something which newcomers are likely to try, and despite the urgings of the experienced gardeners, they often begin planting in May rather than in November. After one or two attempts, they are ready to give up entirely.

Probably the reason why annuals are not very popular in south

Florida is because gardening here is a year around activity. In a land where we step out of doors and encounter plants in flower or in fruit at any day of the year, it is pretty difficult to get excited about a spring flower garden. Gardening in the north, on the other hand, is a spring and summer event. For long months trees are bare of leaves and for long periods the ground is covered with snow. It seems only natural that people living under such conditions should like to plant spring gardens and to tend them, and that they should look forward with enthusiasm to masses of colorful bloom. Perhaps this is one reason why flower shows are so much more popular in the north than they are in south Florida. How can we be enthusiastic about a flower show when upon arrival we find relatively the same plants that we saw lining the streets along the way? Northern gardening is associated primarily with flower production, and northern gardeners, accustomed to growing annuals back home, are likely to want to do the same kind of gardening in south Florida. Some have outstanding success, but for the most part attempts are either failures or the results are mediocre.

Annuals which give good returns for the investment of time and labor necessary to grow them here are the calendula, dianthus, cock's comb, tithonia, gaillardia, marigold, nasturium, petunia, red salavia, balsam, and zinnia. But if planted at the beginning of the cool dry season and not later than the first of January, the gardener should be fairly successful with a wide number of others, including alyssum, bachelor's button, candytuft, cosmos, globe amaranth, hollyhock, lobelia, lupine, pansy, sweet pea, and verbena.

Because of the numerous soil pests which make the growing of annuals from seed difficult, it is always best to buy started plants from garden stores or from nurserymen who specialize in growing them. Moreover, this is a fairly sound way of getting started with the kinds of annuals that do well for the time of season they are planted.

Seeds of annual flowering plants are usually very small, and generally the seeds are planted in seed flats and later the seedling plants are transplanted to garden beds or rows. Exceptions are the nasturtium and sweet pea, which are seeded directly in the garden where they are to grow permanently.

All of the annuals mentioned here should be grown in full sun. They also require very careful attention, especially when the young seedlings are first transplanted to permanent locations. Plants must be kept growing rapidly with the aid of fertilizer and moisture, although it must be remembered that both can be overdone. Any general purpose fertilizer is suitable for mixing with the soil, when preparing rows or beds a week before planting. An application of a soluble or liquid fertilizer is helpful to get young plants off to a fast start. After plants are well established, usually once or twice a week is often enough to water, providing watering is done thoroughly. Thorough watering means wetting the soil to a depth of several inches. Pests must be kept under control. Insecticides should never be applied when the soil is excessively dry.

Most seed companies produce valuable catalogs, which give a great many tips on gardening and also give detailed directions for planting. Seed packages also contain directions. But when buying started plants from nurserymen, the gardener should remember to ask about planting distances and also seek tips on culture. Mistakes are to be expected. But one generally learns more from failures than from success, and the gardener who sticks with annuals and learns how to do a good job of growing them can produce magnificent color in his garden in winter or early spring.

Tropical Perennials

Unlike the hardy northern annuals which can be grown in south Florida during the cool season, the common perennials of northern gardens do not thrive here. Northern bulb plants and other herbaceous perennials do poorly because they have difficulty surviving the long warm humid summers. Only those bulb plants, such as the tulip, which can be pre-chilled under artificial refrigeration, are recommended for planting in south Florida. And these should be planted with full understanding that they will not survive through the summer. Thus the peony, the clematis, anemone, and the delphinium might as well be forgotten unless the gardener can figure some way to provide these cold-tolerant plants with back-home conditions.

We need not be concerned, however, because northern perennials do not thrive in Florida's subtropical climate. We have a large number of tropical perennials which are just as useful and just as colorful and unusual as former northern gardeners were used to back home. Moreover, many of these perennials which do so well in south Florida are widely cultivated as indoor plants in the north. In south Florida many of the perennials are listed under "exotic" plants. Among these are the dracaena, caladium, anthurium, philodendron, and strelitzia. Nor should we forget the several kinds of heliconia, the succulents such as kalanchoe, or the euphorbias, especially *Euphorbia splendens,* or crown-of-thorns. Add to these the many ground covers which have colorful foliage or which produce bright flowers, as well as the many bulb plants, including the amaryllis, the crinum, and the gloriosa lily. Most of the tropical perennials thrive best in older garden where the soil has been improved and mellowed with age, and where there is at least part shade and some protection from cold, drying winds in winter.

While the orchid is never included in the category of perennials, it should not be forgotten that this is one of our outstanding flowering plants in south Florida and deserves consideration whereever color is wanted. Most orchids bloom over a long period, especially the epidendrums and the vandas. *Epidendrum radicans,* although a very common type of orchid, can produce spectacular masses of bright red flowers when plants are massed in ground cover effect. This orchid grows in full sun, and thrives in a deep mulch made of wood chips, tree fern, and a few wood shavings to help hold moisture.

Here is a list of perennial plants recommended for their colorful bloom or for their form or foliage:

AECHMEA. Very colorful bromeliads, of which there are dozens of species, grown for their unusual forms as well as for the bright color of their bloom. Propagated from offsets.

AGAVE. Century plant. There are several species of this large plant which is associated with desert conditions. The agave flowers when it is ten or more years old, sending up a single spike from the center which reaches twenty to thirty feet. After flowering it dies. Propagated from seedlings which volunteer about old plants after flowering.

AGLAONEMA. A common indoor plant, several species of which can grow out of doors in south Florida under suitable shade and wind protection. *Aglaonema roebelinii* is one of the most satisfying of indoor plants, but it is slow in growth and is not widely seen in the nursery trade. The name *Schismatoglottis roebelinii* is preferred by some botanists. Propagated by cuttings of old canes.

ALOCASIA. Elephant-ear plant. Several species are planted in south Florida, including some with giant leaves up to three feet across. Many of the common alocasias can be grown out of doors, and are easily propagated by lifting small plants which grow from tubers at the base of the older plants.

ALOE. One of the most satisfactory ornamental plants for dry, sunny locations. A large number of species and hybrids are available. The unusual flowers, many of them very colorful, are produced on long spikes. Some of them stay in bloom for several weeks. The aloe is one of the best plants available for the boulder garden, where one is trying to simulate desert-like conditions. Seeds or offsets.

ALPINIA NUTANS. Shell ginger. The most common of the gingers in south Florida, producing clusters of irregularly bell-shaped flowers at the tips of banana-like foliage. Propagated from suckers.

ALPINIA PURPURATA. Red ginger. Plant is similar to that of shell ginger, but flower stalks, which stand upright, are bright red. Propagated from suckers.

AMARYLLIS. Belladonna lily. Several varieties of this bulbous plant are grown in Florida. Propagated by dividing old plants in late summer.

ANTHURIUM. A common greenhouse plant, grown rarely in jungle gardens in south Florida. Seeds.

ASPARAGUS. Asparagus fern. At least three species of ornamental asparagus are planted in south Florida. *Asparagus falcatus* is a tall climber, reaching twenty to thirty feet. *Asparagus sprengeri* is occasionally used as a ground cover. Division.

ASPIDISTRA. A plant with tough, leathery leaves growing from thick roots. Difficult to eradicate once it is started. Aspidistra was one time a popular indoor plant. Propagated by division.

BEGONIA. Only a few of the hundreds of varieties of begonias thrive outdoors in south Florida, and they require a fibrous soil which is free from nematodes. Although it is possible to grow begonias in full sun in south Florida, the leaves usually burn in summer if the soil becomes low in moisture. It is best to grow this plant in part shade. Only the rhizomatous type of begonia grows well here. If the bulbous type of begonia is planted in Florida it blooms for one season and declines. One of the outstanding begonias for south Florida is *Begonia nelumbiifolia.* Leaves grow to over a foot in length and almost as wide on well grown plants. Propagated from rhizome cuttings.

BELOPERONE GUTTATA. Shrimp plant. Reddish - brown flower clusters resemble shrimps. A highly desirable plant when well grown. Should be suitable for use as a ground cover under some conditions. Colorful flowers persist for several weeks in spring and summer. Cuttings.

BILLBERGIA. One of the most widely grown and colorful of the bromeliads. *Billbergia pyramidalas,* the most common of the bromeliads planted in south Florida, produces spikes of brilliantly red bracts pushing out of a rosette of deep green leaves. Offsets.

CALADIUM. An endless number of varieties of this colorful plant are available. Grown in south Florida at all times of the year, but leaves may fade or even burn if planted in full sun in summer, unless the soil is kept moist. Division of tubers.

CEREUS. Night flowering cactus. Both a tree form and a climbing form of the night blooming cactus are planted in south Florida. Although the individual flowers are spectacular, they bloom only at night, and the plant itself is often difficult to deal with from a landscape designer's standpoint. The tree form may be fitted into many types of rock gardens, especially where desert like conditions are simulated. Cuttings.

CHRYSANTHEMUM. A difficult plant to grow in south Florida except by the master gardener. A small flowering type, known as the azalea mum, will grow here all year, but usually looks very poor because fungus attacks its leaves. Outstanding chrysanthemums are difficult enough to grow by experts. Those who consider themselves expert gardeners may try rooted cuttings planted in the fall.

COLOCASIA. This is an elephant ear type of plant related to the alocasia. *Colocasia esculenta,* also known as taro, is a common food plant in the tropics, yielding an edible tuber which is usually prepared by baking. It is grown commercially in north Florida to a limited extent. Division of tubers.

CORDYLINE TERMINALIS. This is one of the most colorful plants from the tropics, and a large number of varieties are offered in the nursery trade. Best known of the cordyline is the ti plant, grown for its large, rich green leaves. The leaves of other cordylines may be bright red in color. A wide variety of hues exist, from red to pink and copper, mixed with mottling of white and deep metallic green. Cordylines are propagated by cutting up old canes.

COSTUS. Related to the ginger, this plant bears flowers at the tips of reed-like stems which resemble small pineapples. Suckers.

CRINUM. Among the largest of bulb plants, the crinum is a relative of the amaryllis, and is widely used in landscape plantings here. Its leaves sometimes develop a reddish color caused by a fungus. This can be controlled if a fungicide is applied when the first signs are noted. Division.

DRACAENA. This should probably be classed with the trees and shrubs rather than with the perennials, as some species get to be sixty feet and may grow to be hundreds, or perhaps thousands, of years old. *Dracaena fragrans* is a common variety used in landscape planting, reaching ten to twelve feet in height at times, but which can be controlled by cutting taller canes back. Old canes are easily rooted.

EPIDENDRUM RADICANS. Althought this is an orchid, it is mentioned here because it can be effectively grown in large plant beds. It is a terrestial orchid, growing in humus rather than on tree trunks. It is inexpensive and very easy to propagate from offsets.

EUPHORBIA MILII. *Euphorbia splendens.* Crown-of-thorns. One of the easiest plants to grow and one of the most satisfactory in flowering. The flowers of some varieties are bright red, while others may be salmon pink. This plant prefers a fairly dry and well drained location. Cuttings.

GERBERA JAMESONII. African daisy. A true perennial "daisy" for south Florida, but must be grown in soil which is free from nematodes. Division.

GLADIOLUS. An almost endless number of varieties in various colors are available. Corms, (tubers) must be planted deep enough so that tall plants will not fall over when they begin to flower. Corms are usually planted in trenches about six inches deep and the trenches filled as the plants make their growth. Fertilizer may be placed in the trenches before soil is pulled into them.

GLORIOSA ROTHSCHILDIANA. Several varieties of this climbing lily are available. Tubers are often planted near the base of shrubs so that the lily vine may grow over the shrubs.

GUZMANIA. A colorful bromeliad, of which there are several species and hybrids. Offsets.

HEDYCHIUM. Only one species, *Hedychium coronorium*, the "white ginger," is widely grown in Florida. The plant resembles that of the shell ginger, of which it is a relative. The white flowers are borne in clusters at the tips of the leafy stems four to six feet high. Suckers.

HELICONIA. Numerous species of this banana relative are planted in south Florida, but the heliconia is kept from being widely popular because it is subject to damage by wind. Plants should be grown in protected locations. They are not recommended for young, open gardens. The heliconia is grown for its spectacular inflorescence, many of which are brilliant red. It is often confused with strelitzia. Propagated by division.

HYMENOCALLIS. Spider lily. Resembles the larger crinum, and are usually sold by nurserymen by the name of "crinum lily." Division.

KALANCHOE. Several species of this succulent plant are propagated in Florida, some of which are used as ground covers. See **Ground Covers.**

LILY. Several varieties of Easter lilies are planted in south Florida, the most successful being the Bermuda. Planting is usually done from late fall to mid-winter. If it is desired to have Easter lilies bloom at Easter time, planting should be started about a hundred and thirty days before Easter falls, and additional plantings made every two weeks for six weeks. Division of bulbs.

LIRIOPE MUSCARI. Blue lily turf. A member of the lily family and used to some extent as a ground cover in semi-shaded areas. Does not flower well in south Florida. Division.

MORAEA. An Iris-like plant which produces small yellow flowers. Grown mainly for its evergreen foliage. Plants reach about two feet high when grown under good conditions. Division.

OPUNTIA. Prickly pear cactus. This is a common cactus, with large thick fleshy leaves which are armed with sharp spines. Grown mainly as an oddity, but quite suitable for desert style gardens or seashore areas. Cuttings.

PEDILANTHUS TITHYMALOIDES. Devil's backbone. A fleshy stemmed plant producing large numbers of red tinged flowers throughout much of the year. Not widely propagated because it is so common. Cuttings or division.

PEPEROMIA. Numerous species of this fleshy leafed plant, of which *Peperomia obtusifolia* is the most widely planted outdoors in Florida. A satisfactory plant for use as a ground cover in shady locations. Cuttings.

PHILODENDRON. Although this is a common green house plant, it is widely grown out of doors in south Florida and may be considered one of the herbacious perennials. Both vining and arborescent species are available. Over one hundred species and hybrids are propagated commercially by nurserymen. Cuttings, division, or seeds.

PILEA CADIEREI. Aluminum plant. See **Ground Covers.**

PLATYCERIUM. Stag-horn fern. A popular type of fern, of which several species are sold by nurserymen in south Florida, but planted only in trees or on slabs of moss and other suitable preparations in greenhouses. Never grown in the soil. Offsets.

PORTULACA GRANDIFLORA. Moss rose. A soft brittle stemmed herb producing large numbers of rose, red or yellow flowers. Usually grown as a perennial in south Florida, but its growth is limited because of its susceptibility to nematodes, Cuttings.

POTHOS. See SCINDAPSUS.

ROSE. Most varieties of tea and floribunda roses can be grown satisfactorily in south Florida, provided rich conditions are developed and maintained. Roses must be kept free of insects and diseases. Climbing roses may grow well in south Florida but may not flower. Roses are normally planted in south Florida in the fall, about the first of November. Plants should bloom at all seasons, but the bloom which appears during the cool season is usually the best. Only top quality rose bushes should be planted, as inferior plants do not live very long and produce poor flowers. Local varieties, known as "rock roses," are propagated from cuttings.

SALVIA SPLENDENS. Scarlet sage. Although this plant makes a shrub it is usually cultivated as an annual. Seeds.

SANSEVIERA. Several varieties and interesting forms of this plant are available for ornamental planting. It thrives in shade or sun and can withstand long neglect, except for one reason. Slugs feed on the foliage and render the plants unsightly. Periodic applications of slug bait in plant beds will keep down these pests. Division.

SCINDAPSUS AUREA. Pothos. Devil's ivy. This is a well known fleshy climber, with large heart-shaped leaves. The leaves are usually variegated with green and yellow, while others have streaks of white in the foliage. Cuttings.

SPATHIPHYLLUM. Grown in semi shaded locations for its large deep green leaves and its numerous white blooms, or spathes. Must be protected from slugs. Division.

STAPELIA. Star-fish flower. This plant, often called a cactus, is a succulent which produces brown and yellow star shaped flowers, sometimes up to twelve inches across. Grown mainly as an oddity. Cuttings.

STRELITZIA NICOLAI: This is the larger of the two better known species of strelitzia grown in Florida. It produces a trunk and reaches ten or twelve feet high. Seeds.

STERLITZIA REGINAE. Bird-of-paradise. This is the trunkless form of the strelitzia, and the species which produces the most spectacular flower. Selection for the planting site of the strelitzia should be well chosen so that they do not have to be transplanted. Nor should the strelitzia be dug up and divided, as one would divide an amaryllis. These are slow growing plants and should not be disturbed. Flowering is best on old, crowded specimens. Plant in full sun or in a location where sun will be on plant most of the day. Seeds.

SYNGONIUM. A relative of the philodendron and a climber with much the same habit. Requires shade. Cuttings.

TACCA. This is one of the few plants which produces an almost black flower. The tacca is grown mainly for its large olive-green leaves. Seeds or division.

VINCA ROSEA. Periwinkle. A common sprawling herbacious plant, often seen growing in vacant lots. Two main varieties are seen, one with purple-red flowers and another with white flowers. Usually grown from seeds.

ZEPHYRANTHES. Rain lily. A small bulbous plant with small grass-like leaves. Pink and white flowering varieties are seen. Flowering may take place several times in summer and fall. Division of bulbs.

Giving each species of plant a botanical name is necessary. There are dozens of common names for some plants, in several languages, from Aztec to Zulu. Botanical names are standard throughout the world.

Colorful flowers and unique form make bromeliad desirable

Orchids come in many forms and endless number of colors

Chapter 14

ORCHIDS AND BROMELIADS

The growing of orchids, bromeliads, and fancy greenhouse plants has become increasingly popular in recent years. Many home owners have gone to considerable expense to erect greenhouses for their collections, where both moisture and temperature can be controlled. Others grow their plants in a shade house made of slats or plastic screen. Some are satisfied to grow orchids in trees and they have derived great satisfaction from the growing of philodendrons, anthuriums, and many other jungle-type plants on the ground beneath their trees.

Each of the several approaches to growing these plants has its merits. A number of fancy foliage plants, and a few orchids, require the protection of a glass house where heat can be provided on cold nights. The phalaenopsis orchid and also the popular foliage plant, dieffenbachia, are too tender to cold to be grown outdoors in south Florida during the winter. However, most orchids and foliage plants, as well as just about all of the bromeliads, thrive outdoors the year round in south Florida, under a canopy such as that provided by live oaks or cabbage palms. Broken shade overhead is most important. Windbreaks also must be provided, either a living wall made of shrubbery or a man-made wall made of stone, concrete blocks, wood, or bamboo fence. Cold, dry winds coming out of the northwest or north in the winter time can do severe damage to many ornamentals, and especially such foliage plants as anthuriums and philodendrons, whose leaves may be whipped about as well as dried out by the wind.

A special irrigation system should be provided, in the tree tops as well as on the ground. The tree-top, or upperstory sprinklers, need never be run but a few minutes at a time, to provide a humid condition for orchids, bromeliads, staghorn ferns, or other such upper-story plants which are growing on branches and tree trunks. Plastic pipe may be used for installing these sprinklers. It must be kept in mind, however, that water pressure is rapidly lost as it is being forced up twenty-five to forty feet to a treetop.

Thus little pressure will be available and only light easy-to-operate sprinklers may be used, such as a whirling type.

These sprinklers seldom put out enough water to provide plants growing on the ground with the moisture they need. Thus a second set of sprinklers must be set at about a seven-foot level to provide adequate irrigation of under-story plants. Such a system has been installed in a simulated rain forest at the Fairchild Tropical Garden, and it has worked exceedingly well. The upper-story plants are watered at least once a day, with the sprinklers cut on ten to fifteen minutes at a time. The under-story plants may need water about twice a week, but in order to wet the soil as well as the mulch in which these plants are growing, it is necessary to run the sprinkler thirty minutes or more each time.

While the live oak is probably the best of all jungle garden nurse trees, any trees may be used except those like the ficus or the bischofia which produce a solid canopy of shade, or the sleek gumbo limbo which is constantly shedding its bark in thin layers like onion peel. Nor are such trees recommended which shed their leaves and are bare of foliage for long periods during winter. Among these are the woman's tongue tree, bombax, and poinciana.

Trees, large and small, which may be considered for a jungle garden are calabash *(Crescentia cujete),* West Indian cherry tree, bucida, and most of the palms except the coconut, royal, and any other which sheds large leaves that come banging to the ground with a crash. A colony of Florida's native sabal palms should be rated next to the live oak in providing near ideal conditions for the establishment of a jungle garden. The rough trunk, after the old leaf bases have decayed and dropped, makes the best surface for the planting of orchids. It is not a good idea to plant orchids in the old leaf stem base, as these decay and drop in time, bringing the orchid plant down with them. But these old boots, as they are called, may be good places for the establishment of temporary things, such as climbing ferns or climbing philodendrons. A few trees should be given over to climbers in order to achieve the lush effect associated with the tropical jungle. But climbers should not be started on trees or palms where orchids and bromeliads are established, since the climbers will cover the more desirable plants, shading and choking them out.

In general, old and poorly cared for mango and avocado trees also may be used as jungle garden nurse trees. Young and vigorous mangos or avocados may produce too heavy foliage for the growing of orchids and bromeliads on their branches.

Attaching Orchids to Trees

Several methods are used for attaching orchids and other epiphytic plants to trees. The most common method is the use of galvanized nails and staples. Strips of plastic screen also may be stretched over the plants and nailed to the trunk or a branch. Nails or staples, however, provide invasion routes for fungus, especially in palms. They should not be used if any other method can be found. A better method is to use narrow strips of plastic screen long enough to tie the orchid plants to a trunk or branch. Within a few months the plant has securely attached itself by putting out numerous new roots. The piece of plastic can then be cut loose. No harm is done to the tree. Pieces of old insulated wire will serve about as well. All old roots should be removed from orchid plants before they are attached. Only through the growth of a new root system can the orchid attach itself to a tree. The old roots are of no value anyway and those still living on the plant are likely to die.

For the growing of under-story plants, such as philodendrons, anthuriums, tree ferns, or shade palms such as the chamaedoreas, a well drained soil containing a great amount of humus is needed. If the soil is poor and sandy, two inches of peat moss may be scattered over the sand and incorporated with the aid of a fork or a shovel. To insure good drainage and aeration, an inch of gravel might be included with the peat moss. Limestone gravel is quite suitable, and is much less expensive than the brown river gravel sometimes used for this purpose. The gravel can be raked lightly into the surface. On top of this well aerated humus base, several inches of mulching material should be added, leaves, grass clippings, wood chips, saw dust, or wood shavings. If peat moss must be used as mulch it should be mixed thoroughly with gravel, and perhaps one-half to one inch of coarse perlite. The idea is to develop a planting base similar to the floor of a tropical forest. Good aeration is very important.

If the soil is shallow, and only an inch or so of sand lies over limestone rock, it may be necessary to add two or three inches of sand in order to make a suitable place for growing jungle plants. Use a coarse builder's sand, and never use muck or marl. Gravel and peat moss can be mixed with the sand at the time it is spread.

Planting is done directly into whatever base is provided. A fertilizer made up of half sheep manure and regular garden fertilizer can be mixed with the soil when a plant is set out. No large amount need be applied, but a light application of fertilizer should be made once a month until plants are thoroughly established.

Continual renewal of mulch is necessary, as organic matter disappears rapidly. Leaves may lie where they fall, and if additional leaves can be obtained they should be spread about the plants.

Many kinds of orchids are suitable for growing on trees. The genus *Cattleya* is the best known. There are many species of cattleyas and thousands of hybrid cattleyas. The common species and the older varieties, much less expensive than the modern hybrids, should be used for trees. Expensive plants should be kept in containers, and in a more protected location than a tree top. Varieties can be selected which bloom at different times of the year, and thus the flowering season for the jungle orchid collection can be spread out through most of the year.

There are many other kinds of orchids which are just as interesting as the cattleya. These include the genera *Oncidium, Epidendrum, Dendrobium, Brassavola,* and *Laelia.*

The best way to learn orchids is to study them as you buy, and try to learn all the cultural needs of each kind as you acquire it. There are many private and commercial collections of orchids in Florida which are grown on trees under jungle-like conditions. These areas should be studied and methods of culture noted. Many books are available on orchid culture. Very often a jungle garden of orchids can be developed in conjunction with a greenhouse type of orchid collection. This is a very satisfactory arrangement for those who develop orchid growing into an untiring hobby.

Bromeliads

There are many other tropical plant groups which are just as attractive as orchids, at least to many persons. In the bromeliad group, for instance, there are hundreds of species of most unusual plant forms, which produce as colorful bloom as your finest orchids. Moreover, this is a group of plants which is not difficult to grow, and may be grown successfully by any amateur, and with a great deal of satisfaction. Many of these plants grow as well in trees as on the forest floor. Most of them require about the same amount of shade as an orchid, and do quite well in the same jungle garden in which the orchids are grown. On the jungle floor, these plants must have excellent drainage, and thus extra amounts of both gravel and perlite might well be used when preparing the soil for them.

The ornamental bromeliads are in the same family with the pineapple and Spanish moss. They come from the tropics, but are well adapted to south Florida's climate. They vary greatly in their forms, colors, and unusual markings, as well as in their brilliant and sometimes odd flowers. These plants are shaped so that they hold water in their leaves, and one doesn't have to worry too much about their drying out and dying of neglect.

Most of the bromeliads resemble a pineapple plant to some extent, but one really has to see these plants to appreciate them. However, the bromeliads have not become as popular in this country as they have in Europe, where they are more common as house plants.

There are several genera of the bromeliads, including *Aechmea, Billbergia, Cryptanthus, Guzmania, Neoregelia,* and *Vriesia.* Among these are many species. In recent years, bromeliad growers have been adding a number of hybrids each year. Flowers last for weeks.

The bromeliads thrive on the same general treatment as orchids, including periodic applications of soluble fertilizer. Malathion is favored for control of scale insects which sometimes get on bromeliads, just as this insecticide is useful for the control of scales on orchid plants.

There is one very important precaution to be remembered in connection with the growing of bromeliads. Copper in any form is deadly to them. A copper fungicide, if used for fungus control in a greenhouse or in a jungle where bromeliads are growing, will almost invariably result in severe injury or death of all the bromeliads with which it comes in contact. Copper spray is specific for the control of Spanish moss in pecan and live oak trees throughout the southeastern states where it grows.

In the planning and planting of a jungle garden, design should be considered just as in any other kind. Space, mass, texture, and color are important here. There must be background plants, middle-ground plants, and foreground plants. There must be vistas and open areas. Large shrubs or cluster palms may be planted along the back or borders of the jungle gardens to provide background. A jungle garden is hardly satisfactory if one is distracted by the flapping in the wind of a neighbor's wash while one tries to enjoy the bloom of a few orchid plants or a collection of philodendrons.

Colorful plants in general do not fit well into the jungle garden. Large amounts of color are out of place here. Mass and texture should be the dominating features, with blooms of orchids, bromeliads, and anthuriums appearing here and there as special attractions. These individual colors should not have to compete with masses of hibiscus and crotons. Large philodendrons should be planted along the borders or in the background, and smaller ones in the foreground. Nor should heavy climbers be placed in the foreground. The large-leafed monstera and raphidophora should be reserved for the sides or to the back.

There's nothing wrong with using large boulders in the jungle garden. These may serve for philodendrons to climb over, or even for ferns. If the surface of the boulders is rough enough, bromeliads may be trained to grow on them. If kept moist, many types of mosses and epiphytes will grow on the surface. While pools also may be built in jungle gardens, it should be remembered that water lilies do not bloom when grown in the shade.

There are both advantages and disadvantages to the growing of plants under jungle-like conditions. Several types of pests may damage the plants, including scales, cockroaches, and slugs. Caterpillars or grasshopers will sometimes take a liking to the finest orchid bloom. Squirrels sometimes damage plants. Moreover, there is the thief to contend with. Where one uses valuable plants in such a collection it usually is necessary to erect a thief-proof fence around the place. However, if one is satisfied with the ordinary orchids, bromeliads, and philodendron, with a sprinkling of choice varieties, there should be very little danger of thievery. It is when one builds a collection of rare things and advertises the collection by inviting large numbers of people to inspect it that the danger of theft increases.

Most insects are harmless, and many are beneficial. Even the termite helps to disintegrate fallen trees in the forest. Predatory insects live on other insects, especially on those which consume plants.

Coccothrinax argentata is a tough palm for the windy seashore

Plants that can take salt spray should be set on bayside

Chapter 15

SPECIAL GARDENS

1. The Seashore Garden

If you should build a cottage on the seashore, treat with a great amount of respect the lowly scrub plants you find growing there. If you already have built on the seashore, after first destroying those "ugly, scrawny things" along the strand, you face the task of finding other plants just as rugged as those you have removed. One thing is certain: you can't grow Christmas trees on the beach. No matter how tough a plant is, or how resistant it may be to salt spray and the buffeting of the wind and flying sand, you can never train a plant growing on the seashore into the perfect form you may desire. On the beach it is the elements which do the training, not the gardener.

The best thing for the beach gardener to do then, is to work with the elements, not against them. Plants should be selected with the full knowledge that they will have some hard going, that they will become gnarled, and one-sided, and that the foliage even on the toughest species may from time to time be burned by salt spray and sand. Here is no place for the formal type of garden. One should think in terms of a seashore garden, of a garden that looks like the seashore, rugged and windswept.

The southeast Florida seashore is not an easy place to get a plant established. The soil is mainly of sand and broken shell matter. It contains little humus or nutrients. It is highly alkaline. Moreover, the sand dries out rapidly and newly set plants will surely suffer if watering is neglected. It is helpful to incorporate peat moss and organic fertilizer such as sheep manure or sewage sludge with the soil before planting. Once a plant is established, a regular garden type fertilizer may be used.

An ecological study of any exposed seashore, such as along the east coast of Florida, invariably will reveal a rapid change in plant species as one walks with one's back to the sea. The first line of plants may be sea grapes, yucca, sea oats, scaveola, sea lavender, or, by chance, perhaps a coconut palm. Behind this protective front line planting one finds the saw palmetto, taller sea grape plants, pigeon plums, wax myrtle, prickly pear cactus, the coco-plum, and still further back taller specimens of all of these, plus many native hammock trees, including the live oak. But when one gets a hundred yards from the front line planting along the edge of the beach, the gradual rise of the almost solid protective screen makes it possible for many kinds of plants to grow, which, if they by chance sprouted by the seashore, would be quickly killed through exposure to the elements. This gives one a good hint of how important protection is, and how it is possible to develop fair protection only a few feet back from the beach.

A surprising amount of protection may be obtained by building a low retaining wall near the edge of the beach. This is especially valuable for breaking the force of the waves which are tossed up on the beach during storms. We make no reference to the effect of hurricane winds on beach-front gardens. There is no way to protect a seashore garden from the fury of a hurricane. Very large boulders may also give some security to a seashore garden, but the boulders should weight a ton or more each, so that hurricane waves cannot send them crashing into the house.

Probably the best of all front line seashore plants is the common yucca, or Spanish bayonet. If one wants to soften the exposed side of the yucca, some of the native fleshy leafed plants found growing along the seashore may be transplanted. Many of these are difficult to transplant because of the long tap roots they develop. Only the younger plants normally can be transplanted successfully. It is difficult to move an old specimen of sea lavender or scaveola.

Immediately behind the protected yuccas, several shrubby specimens of sea grapes might be placed as further protection against wind and salt spray. Behind the sea grapes one can then feel free to plant a wide variety less tolerant to salt spray, including carissa, jacquinia, ochrosia, ligustrum, pittosporum, pandanus, raphiolepsis, as well as such ground covers as lantana, wedelia, and several kinds of kalanchoe. The aloe also may be planted directly on the edge of the beach, in the face of flying salt spray, but these plants look their best where they get some shelter from the wind, especially when they send up their spikes of flowers. Thus, they serve especially for ground cover in protected areas.

The beach front garden should be designed to give the best effect of the beach and the sea viewed from the house. Most plantings should be limited to the borders of the garden. Only very low growing plants or tall palms should be planted directly in front of the house. Nothing should be done to spoil the view. The aim of the design should be to enhance the view of the ocean. For this purpose, no other plant exceeds the coconut palm, which is naturally assoicated with the sea shore. Its large leaves are designed to give freely with the wind, and yet it is always graceful and picturesque. Should it be battered by a storm it can be expected to recover quickly. Actually, the coconut palm is the only one which can be expected to thrive right out on the edge of the beach. Seldom can either the cabbage palm or the saw palmetto exist here, unless they happen to be on rather high sand dunes.

No mention has been made of the Australian pine. Under some conditions this tree undoubtedly should have a place in the seashore landscape. It will grow to the edge of the beach, along with the coconut and the sea grape. However, on the average ocean front lot the Australina pine is objectionable because of the way it cuts off one's view of the horizon.

Along the bay shore, and also along much of the protected coast of the Gulf of Mexico, the situation is vastly different. Except in the most exposed locations one does not have to worry too much about the closeness of salt water. Hibiscus grows along the borders of Biscayne Bay without any injury except during a storm. Many of southeast Florida's finest royal palms grow near the bay's edge. However, as one goes southward along the bay shore, the bay becomes more open than it is in the northern part, and plants need more protection. One has to judge the situation from what neighboring homeowners have already done, unless one happens to be the first person to build a house in an area.

2. The Water Garden

Gardening can be done in water, too. Water in itself is interesting, whether it is in a small pool, a natural pond, a stream or a lake. Moreover, water can be planted.

No other ornamentals in Florida, including the most colorful orchids, make a greater show than a planting of tropical water lilies. Colors come in white, lavender, pink, red, yellow, or blue. These lilies cross readily and many hundreds of outstanding varieties have been developed. A well grown tropical water lily may have one or more flowers in bloom nearly every day in the year.

Both day-blooming and night-blooming varieties are available. The night-blooming water lily, although opening at night, remains in bloom until nearly noon the following day. The day-blooming water lily begins opening of a morning and stays open until evening. Only the tropical variety should be planted in south Florida or in other tropical areas. Even in the north, the tropical water lilies are usually preferred over the hardy types. In cooler climates the tropical water lilies are taken up in the fall, and the bulbs are saved until the next spring, when they again are planted out of doors.

Gardening with water lilies is easy, if one has a pool large enough. Even large wash tubs are sometimes used successfully for growing them. When planted in a pool, and grown under very rich conditions, a single vigorous water lily plant may cover as much as fifty square feet of water surface. If grown in a tub, the plant simply remains smaller and flowers are consequently smaller. If a plant receives but little fertilizer it will tend to develop small leaves and small flowers.

With the idea that water lilies may be grown in a container as small as a wash tub, it seems only natural that local landscape architects might design containers in interesting forms for patios, for the purpose of growing these colorful plants. Italian landscape designers already are doing this. Such containers should be four to five feet in diameter, and should be made so that a planting box might be covered by about ten to twelve inches of water. A twelve-inch azalea clay pot makes an excellent soil container for growing water lilies. When a water lily pool is made, a special receptacle in the bottom should be provided so that the azalea pot might fit snugly into it. Where large pools are made several of these receptacles might be built in, to be used when desired.

In nature, water lilies grow at various depths, from a few inches to as much as five or six feet below the surface. They are capable of adjusting themselves to the rise or fall of water, at least to a limited extent. The plants are rooted on the bottom, and never float on top except when some accident happens to loosen them. The floating leaves are borne on the ends of long petioles. Propagation generally is from divisions of old clumps. Tubers may be ordered from nurserymen specializing in the sale of water lilies. Tubers are planted so that they lie horizontally in the soil rather than upright. Older plants produce seeds and soon the bottom of a pond may be covered with scores of small plants.

A soil mixture containing one part marl and three parts of sandy soil makes a good medium for the growing of water lilies. Add to this a coffee can full of sheep or cow manure for each azalea pot of mixture used, plus the equivalent of a couple of tablespoons of ordinary garden fertilizer. After being thoroughly mixed, the medium is placed in an azalea pot or similar container to within an inch or an inch and a half of the top. The bulb or plant, whichever is used, is now planted, and the rest of the pot is filled with coarse clean sand.

The next step is to sink the pot into water, slowly enough so that by the time the pot is submerged the soil and sand are thoroughly saturated with water. So long as bubbles rise to the surface the soil isn't thoroughly wet. If a pot of soil is sunk into a pool before it is wet, the soil is likely to float to the top.

Keeping water lilies well fed is important, otherwise bloom may be sparse and small. Use a highly organic fertilizer, such as a rose fertilizer, if a special mixture for water lilies is not readily available. Put the equivalent of two tablespoonfuls in a plastic bag, close the end firmly with a rubber band and bury in the soil of the container. Be sure all the air is out of the bag before securing it. It is best to moisten the contents first. Bury in the soil among the roots of the plant. Repeat this about every two months.

Many gardening encyclopedias and gardening periodicals have plans for the building of lily pools. To build a large pool is not an easy job. A lily pool also can be a hazard where there are small children. Toddlers have drowned in them. A pool, however, need not be more than twelve inches deep if azalea pots can be sunk in fixed receptacles.

Lily pools should be stocked with small fish or minnows to keep down mosquito larvae. Gold fish are not recommended, since they burrow into the soil of containers and loosen lily plants, causing them to float to the surface. A lily pool must be built in a sunny location, as the plants do not thrive in shade.

Natural ponds or lakes give one more freedom in the use of lilies and other aquatic plants, provided the water is not so deep that the plants cannot be anchored. Most bodies of water are shallow enough along the shore, however, and lilies may be planted here. Shallow, boggy ponds are the best. In Florida many of these shallow ponds have boggy shores, and they already may be partly filled with aquatic plants.

Florida abounds in many kinds of interesting bog plants, which may be transplanted from the wilds to one's own pond. Some of these plants, however, are quick to take over a pond, and may soon increase until they are out of control. Once a solid mass of bog plants such as pontederia or arrowhead is established, they are very difficult to eradicate without the assistance of power equipment, such as a dragline. The cattail is another plant to avoid. It can be controlled, however, by spraying with the weed killer dalapon.

Aquatic plants are not recommended for planting in fish ponds if a balance between small fry and larger fish is sought through a program of combining fishing and fertilizing. Water lilies and other aquatic plants provide hiding places for the small fish, protecting them from becoming food for the bigger fish. This results in an unbalanced situation. The pond becomes filled with too many small fish, not big enough for the frying pan, and the competition among these small fish for food prevents them from growing. Anyone wishing to develop a balanced situation in a fish pond should consult the Florida Game and Fresh Water Fish Commission in Tallahassee, which not only can supply information but assist in procuring the recommended types of fish.

3. The Patio Garden

The patio garden came from Moorish Africa by way of Spain. Over the centuries, however, the patio has undergone a great many changes. In its earliest conception, the patio was simply a courtyard about which a fortress-like home was built. The Spanish brought the courtyard idea with them to America where, with minor changes, this kind of construction was continued until very recently,

If you have visited older communities in Latin America you may have seen solid walls of homes along the sidewalks on either side of the street. If you had entered the door in one of these walls you would have found yourself in a home surrounding a courtyard, or patio. The patio is invariably surrounded on at least three sides by a loggia, behind which are rooms of the house – the living room, bedrooms, dining room, and kitchen as well as servants' quarters. In larger, more elaborate places, the loggia extends around the four sides, but ordinarily the courtyard ends at an eight-foot wall in the rear. Many modifications of this plan exists, because these patio gardens and homes have been built at different times over several centuries, with attending changes of styles and ideas. Today in Latin America homes are built more or less like homes in the Miami area. Gardens are outdoors instead of indoors, so to speak, surrounding the house instead of being surrounded by the house. But patios still remain popular throughout Latin America and few homes of any size are without them.

Today, in Latin America or in Florida, the patio may be any paved area out of doors, while the design and construction varies from home to home, according to living habit, taste, and income.

Patios are made of many kinds of materials, including loose or frozen gravel, flagstone, brick, key stone, solidly poured concrete, or of terrazzo. The forms of patio terraces vary according to the whims and ability of the designer. Some are simply rectangular, or they may be irregularly laid out with small or large stepping stones. Some are laid out in free-line form. Screening of patios against sand flies and mosquitoes is often necessary.

Occasionally one sees a patio where plants are limited to two or three tubs containing philodendrons or palms. But plants usually play a major role in most patio plans and their selection and care is very important. If shade-loving tropical plants are to be used partial shade is necessary. Only a sun-tolerant plant should be used where patios are exposed to the full force of the sun's rays. It should be realized that a great amount of heat is reflected from patio surfaces and from nearby walls. Shade is especially necessary during the summer. Large areas of paving become very hot and may be unusable except in the evenings. But how to shade large areas of paving can be a difficult problem, unless plastic screen is used.

A practical way of handling the patio shade problem is to provide shade over portions of it, leaving the rest in full sun. Palms are widely used, especially the native cabbage palm. The coconut palm becomes a hazard when planted about the patio. Not only are the nuts themselves dangerous, but the large fronds are capable of doing considerable damage to patio furniture, plants, or even to people, when they fall.

The cabbage palm is close to ideal. Moreover, it is slow growing. A ten-foot palm ordinarily will not grow six inches a year. Cabbage palm leaves are not shed until long after they are dead and have dried. The disadvantage of the cabbage palm is that it produces large clusters of flowers and fruits, and shedding of these may be messy. This problem can be solved by removing the flower clusters as soon as the creamy white bloom begins to fade.

To find a palm or a tree which does not shed leaves or flowers or fruits would be impossible. Some trees are more messy than others. Those which drop fruit, such as the jambolan plum, rose apple, guava, or sapodilla, should be avoided. Neither are trees recommended such as the woman's tongue (*Albizzia lebbeck*), which is bare of foliage for several weeks during the winter and spring, and which sheds a shower of its dried seed pods every time there is a breeze. Both the bucida and the live oak are excellent patio trees, even though they do shed their foliage. But either of these trees might be a disappointment if their branches extend over a terrazza terrace, as the terrazza is sure to be stained from the falling leaves.

The Design Should Fit

The approach to patio design and construction should be very informal if it is to be made beneath large, spreading trees, where there is sure to be an almost constant dropping of leaves, flowers, and fruits. Planting along the borders should be of shrubs and palms. The general atmosphere should be such that the sweeping of leaves off the paved surface and underneath the shrubs and palms will seem natural.

Elaborate paving and construction may require the services of a landscape architect, as well as having the work done by a building contractor. However, the homeowner can build his own patio of brick or stepping stones at very little expense. The area to be paved is cleared, excavated if necessary, and then leveled. A layer of sand is then put down and leveled and the bricks or stepping stones placed on top of this base. Actually, a terrace should not be perfectly level, but should have enough difference in elevation so that rain water will run off in the desired direction. A difference of an inch in ten feet should be sufficient. One should not expect water to drain efficiently between the brick or stepping stones. There isn't enough surface space for the water to run, and, moreover, if the sand beneath tends to pack, there may be very little drainage if at all. Large pools of water may stand over the patio for hours at a time. If cement is used between the bricks or stepping stones, obviously allowance must be made for the water to drain off to one side of the terrace.

In planning a terrace the homeowner should consult various books on gardening architecture, especially the several publications on modern patios, and study methods of construction. Suggested patterns for laying brick also will be found in these publications. Older books on gardening architecture will include many fancy and complicated patterns, which are not found in modern patio design books. These complicated patterns, as well as other kinds of fussy gardening architecture, including statues of nymphs and cherubs, have no place in modern gardening.

Various methods of fitting the brick or stepping stones together may suggest themselves. Some like to use grass between stepping stones, especially the fine bladed *Zoysia tenuifolia.* This grass creates an interesting green pattern in the patio, but it does require some work. Occasionally it must be hand-clipped. It also is subject to damage by sod webworms and one must be prepared to apply an insecticide if damage is noted. If mortar or cement is to be used between the bricks or stepping stones, it is best to apply these materials in a dry state, then moisten slightly so that they will set. It is very difficult to apply mortar or concrete between bricks or stepping stones when wet without making a mess. Another material which is sometimes used is asphalt, especially between stepping stones. The black lines create an interesting pattern. Dry-mixed materials can be purchased from building supply firms.

Since the patio is really an outdoor living area, some method of providing privacy must be considered. A number of methods suggest themselves, including walls, woven redwood fences, board fences, bamboo fences, masses of shrubbery, or clusters of palms set closely together. Wood, unless it is treated, will last but a few years, especially if it comes in contact with the soil. Concrete block walls are expensive to build. First a foundation must be poured and a cap must be poured on top of the block. Moreover, in most municipalities it is not possible to obtain a building permit to erect a wall higher than five feet. Neighbors also may object to high hedges extending the full length of a property line, but are less likely to object to masses of shrubbery or palms planted in strategic locations for screening purposes. Such plants should be tough and capable of withstanding all sorts of conditions, including cold north and northwest winds in winter. Suggested plants for screening are cattley guava, ligustrum, surinam cherry, pigeon plum, orange jasmine, aralia, and three palms, *Ptychosperma macarthuri*, *Chrysalidocarpus lutescens*, and *Rhapis excelsa*.

Sometimes screening plants may be permitted to grow in the form of a small tree, providing shade for philodendrons, and shade loving chamaedorea palms, as well as for a number of interesting ground covers. Rich and interesting tropical backgrounds can be created by such method of planting.

Garden Lighting

Outdoor lighting is especially important about the patio. The lighting should be arranged so that it can be enjoyed from indoors as well as outdoors, for there are many times when the patio cannot be used because of rain or other inclement weather. With proper use of lighting the patio may be a source of pleasure although it is visual. Thus it should be possible to control the patio lights from an indoor switch.

Proper installation of patio lighting is important. A basic system of wiring should be planned, and wires should be run through protective metal conduits. It is best to have the fixtures portable, so they can be changed when necessary. For plants change as they grow, and lighting fixtures often must be changed to be effective. Thus the portable, plug-in type of fixture is usually better than the permanently installed type of fixture. Several protected outlets should be available so that one has lots of freedom in changing garden lighting.

It is dangerous to handle electrical fixtures when they are wet. If it is necessary to handle fixtures just after a rain the current should be cut off first.

High wattage bulbs are not necessary in most fixtures used in garden lighting. Where activities call for the use of more powerful or flood lighting, these may be set up temporarily or else they should be set up so they can be cut on by separate switches. Garden lights should be arranged so as to show up the character of the garden to best advantage. Flood lighting is not needed, and may even be detrimental. Firms selling outdoor lighting fixtures publish catalogs which not only illustrate the many kinds of fixtures available, but offer suggestions for using them.

Generally, the main conduit and outlets for garden lighting should be installed before patio paving is done. To wait until after paving may necessitate digging trenches around the perimeter of the paved area, resulting in more expense in labor, as well as the cost of conduit and wire.

Certainly no patio should be planned without provisions made for lighting it. But modern garden lighting isn't just a practical means of providing illumination for the outdoors. As every landscape architect knows who has to face the problem of placing lights in a well planned garden, this can be a difficult form of art. The results, however, can be tremendously pleasing. Unusual plant forms can be illuminated with sometimes bizarre effect through the use of low-powered spot lights. But however lighting is used to show up plant foliage or plant forms, the result is an effect greatly different from what one can obtain during daylight. Thus, with well designed garden lighting, one creates two gardens in the same location, a day-time garden and a night-time garden, and they are just as distinct as if they were in different worlds.

4. The Vegetable Garden

Vegetable growing is often tackled with enthusiasm by newcomers to Florida, especially by those who think that all one has to do is to drop seeds in a furrow and to cover them. Then the gardener sits back and waits for the expected rapid growth of his plants to harvest size. Such gardeners harvest only disappointment. Vegetable growing in south Florida is not easy, even under the best of conditions. On the average city lot conditions are far from ideal. Good soil, ample sunlight, and plenty of moisture are needed to produce even fair quality vegetables. Some experience in the growing of things is necessary, because the gardener will have to use fertilizer to enrich the soil, and he will have to spray for insects and plant diseases. Garden crops also must be planted in the right season.

Spring opens the gardening season in the north, but in south Florida the vegetable garden ordinarily is started in the fall. While it is possible to have a vegetable garden here all year, only a few crops, such as okra, sweet potatoes, squash, cucumbers, and cantaloupes can be grown in the summer garden. The last three require considerable care for they must be protected from both fungus diseases during rainy weather and from hordes of insects that are likely to prevent the gardener from harvesting a single fruit if protection is not provided.

Usually planted in fall, beginning about the first of October, are broccoli, cabbage, carrots, beets, tomatoes, and peppers. Just about all the vegetable crops grown in the north during the spring and summer can be grown here during the winter. Only the inexperienced gardener, however, is likely to plant a large number of varieties of vegetables. The experienced gardener plants only those varieties which, when harvested from his garden, are superior to the vegetables he can buy in the store. The best examples are tomatoes and sweetcorn. But most persons lack the room to grow sweetcorn, and so we often find gardening efforts confined to the growing of tomatoes. As everyone knows who has sampled home-grown tomatoes, they are much superior to those which can be bought. And of all the vegetables which can be planted in the dooryard garden, hardly any equal the tomato in paying rich dividends for the investment of time and effort in their growing.

Starting Plants

Half a dozen plants of well grown tomatoes can supply a family with all the fruit it can use. Planting should be done in soil which is free from nematodes, because tomatoes are subject to rootknot and do not thrive in soil infested by these pests. If the roots of tomato plants do become infested by nematodes the soil should be treated with a nematicide. The gardener will do well to buy started plants from a nurseryman or garden store rather than try to grow his own from seeds, unless he is interested in a special variety. Most kinds, however, can be grown in Florida and they produce well enough to satisfy the home gardener's needs. Usually the nurseryman growing plants for sale will carry the varieties about which he receives the most favorable comment from those who buy them.

There are several possible approaches to the growing of tomatoes. They can be grown successfully in five-gallon cans, in apple baskets filled with clean wood shavings, in marl soil, in pure sand or in mixtures of sand and peat moss, with nutrients supplied by a soluble fertilizer. Mainly a tomato plant needs to have good drainage and aeration, and it should be planted where it receives sun most of the day. It also requires a constant source of moisture. The tomato is a pig for fertilizer. The gardener must remember that it is not only a fast grower, but cannot be fertilized only a couple of times a year as one would fertilize a tree. It must have a constant source of plant food. A tomato plant's roots grow rapidly, spreading out in the soil in all directions. Fertilizer should be placed just ahead of these roots, so that they will be constantly growing into a new supply.

The fertilizer is, of course, covered. It is dissolved by the soil moisture and becomes distributed in the immediate area of the soil by the time the roots grow into the area. Strong applications of fertilizer made directly to plants can burn them, and the novice gardener must be alert to this danger. Fertilizer applications should be small at first, but made progressively larger as plants grow. The roots of a healthy tomato plant extend much further than most gardeners realize. By the time a plant has reached blooming size, fertilizer should be applied eighteen to twenty-four inches from the plant. But if the gardener is careful to distribute the fertilizer evenly, it may be broadcast over the soil to within a few inches of the plant, and then watered into the soil with a garden hose. Fertilizer can be applied every two weeks.

Once a week is enough to water during dry periods, if it is done thoroughly. The soil should be soaked to several inches deep. Daily sprinkling of plants is not recommended, as it encourages diseases. Keeping plants too wet also may prevent the flowers from setting fruit. While young plants are getting established, however, daily watering may be necessary until the shock of transplanting has been overcome.

Spraying Necessary

Blight, a fungus disease which is capable of killing a tomato plant in about a week, is one of the greatest hazards to the growing of this vegetable in south Florida. The trouble starts in small spots on leaves or stems, as if sprayed with scalding water. In a few days the spots have turned black and have spread throughout the plants. Weekly coverage with a fungicide is necessary to prevent blight. Several fungicides are used successfully including zineb, nabam, manzate, and maneb. Both upper and lower surfaces of leaves must be covered, as well as stems and fruit. A pressure type sprayer is recommended. Spraying is done as a preventive. It is begun as soon as the plants are set out, and is continued until the last of the fruit is harvested. Waiting until the disease gets started is usually too late, as it is very hard to control at this time. The aim is to keep the plants spotless.

Another disease, rust, attacks the lower leaves of tomatoes if they are subjected to a prolonged rainy spell, or when they are watered several times a week under a sprinkler system. Fungicides do not control this disease, as it is caused by a bacteria. It must be controlled by spraying with an antibiotic.

Tomato growing is an experience which can bring either a great deal of pleasure or complete disappointment. The outcome usually depends on the willingness of the gardener to follow through with a program of spraying, fertilizing, and watering. Tomato plants require constant checking for signs of insect invasions. Sometimes slugs attack plants. These should be controlled by applying anti-slug bait. If caterpillars are present, they can be controlled by spraying with an insecticide recommended for tomatoes. As the fruit begins to form sometimes a small worm will begin feeding on them, making small holes and causing the fruits to decay.

The tomato plant can be grown on a trellis, it can be staked, or it can be permitted to sprawl on the ground, like a viney shrub. The use of a trellis or stake is the better method for the home gardener. On a trellis where there is plenty of room a well grown plant may grow several feet tall and cover a large area. It must, of course, be trained and tied to the trellis or stake. Staked tomato vines are usually kept fairly small, and are confined to a single stem. This is done by removing all suckers that appear in the leaf axils. One must be very careful here, because both the flower hand and a sucker grow from the same axil. If tomatoes are to be staked they can be planted fairly close together, within fifteen inches apart, in rows thirty inches apart. Fewer tomatoes will be produced per plant than when grown on a trellis or permitted to sprawl on the ground because the staked plants will be smaller. But the staked tomato plant has advantages over the larger plants. It is much easier to spray and the fruits are likely to be of higher quality.

Growing Sweetcorn

Few gardeners ever have an opportunity to grow sweetcorn. Only those who have plenty of space and good soil can hope to grow sweetcorn successfully. However, the quality of home grown sweetcorn can be so much superior to that of the best varieties purchased at the store that the gardener who is able to produce his own will always remember the experience, if he never grows but a single crop. The main problem is the amount of space required.

In a glade, where the soil can be depended on to hold moisture, plants can be grown six to eight inches apart, in rows about thirty inches apart. In higher areas, where the soil is not only poor but where so much of the moisture must be supplied through irrigation, corn plants should stand no closer than twelve inches apart in the rows. Since the best plants can be depended upon to produce no more than two ears of corn, considerable space will be required. The city lot garden simply does not have the space or conditions for the growing of sweetcorn, or for most other vegetables for that matter.

Two outstanding varieties, Golden Bantam and Seneca Chief, are recommended for the home gardener. Usually Golden Bantam can be purchased in seed stores. Seneca Chief may have to be ordered from the Robson Seed Farms, Hall, N. Y.

Sweetcorn must be grown under the richest of conditions, a reason why it is so popular in the rich black soil of the Everglades, where more than thirty thousand acres are planted each year. The home gardener must supply this needed richness through applications of high nitrogen fertilizer. The fertilizer is applied on the sides of the row, as for the tomato. Applications can be made every two weeks. Corn also must be kept well irrigated, but daily watering can be more harmful than helpful. Once or twice a week is usually sufficient, if the soil is thoroughly moistened.

Because of fungus diseases which attack sweetcorn, it can be successfully grown only during the coolest months. The first crop can be planted about the middle of October and the last planting should be made no later than the first week in January. Two insects stand between the gardener and the successful harvest of his crop. They are the corn earworm and the corn silkworm. When buying seed corn, the gardener might as well buy a quantity of Sevin or DDT for control of these pests. As soon as the corn has sprouted, begin applying the insecticide weekly, covering the entire plant, and especially the bud. Continue the weekly applications until the first silks appear on the new ears. From then until the time the silks have died applications must be made every third day.

All the troubles one has to go to in order to grow sweetcorn successfully are likely to cause one to wonder if it is worth the effort. Few gardeners try it more than once or twice, but the growing of any crop successfully, and being able to harvest and to eat it, is a worthwhile experience.

The growing of vegetables should be a good hobby for a retired person with time on his hands. If space is available, and not much space is needed if one does not try to grow such crops as sweetcorn, good soil can be brought in. Many very successful gardens have been made by inclosing a twenty by twenty foot area with thin concrete blocks and putting down six inches of marl and humus such as muck, or black sandy soil. The humus will disappear after a few months but it can be replaced each year by digging in an inch or two of peat moss. If the area is not to be used during the summer for the growing of vegetables, it can be planted in a cover crop such as crotalaria, which will help to keep down nematodes. These pests are likely to increase rapidly if such garden plots are permitted to grow up in weeds. The soil also becomes full of weed seeds, too, if weeds are permitted to grow, and their sprouting during the part of the year the area is used for vegetable growing adds a great deal of unnecessary work for the gardener.

Recommended crops for such a garden plot include pole beans, cabbage, carrots, leaf lettuce, peppers, beets, broccoli, radishes, tomatoes, and turnips. There is usually not enough space for the growing of melon crops, and one is not likely to grow lima beans or English peas which are superior to those one buys. It is also doubtful whether the home gardener will grow superior beets or carrots. Most certainly he should not try to grow head lettuce, cauliflower, or celery. Bush beans make an excellent crop for the small garden, if one uses the high quality varieties recommended for the home garden and avoids commercial varieties planted by farmers.

Strawberries in Winter

The strawberry does well in south Florida, but usually requires more space than can be spared in the small home garden. Northern varieties do poorly here. Only the variety Florida Ninety should be planted. Close to one hundred plants will be needed to yield between a pint and a quart of berries a day. Plants are available after the middle of October and should be planted as soon as they can be obtained. Set Florida Ninety at least twelve inches apart, in rows two feet apart. Like other crops, strawberries require fertilizer and moisture. It is a good idea to start off by digging in five pounds of a high grade garden fertilizer to each one hundred square feet of space. Additional fertilizer can be applied, farther and farther from the plants as the roots extend into the row middles.

Newly set out plants should be watered daily for the first week. A soluble fertilizer may be added to the first irrigation applied after the plants are set out. The same garden type fertilizer applied when the soil was prepared may be applied again at the end of the first two or three weeks, and at six week intervals thereafter.

It is important that plants be set out correctly. The buds must not be covered, nor should the plant be set so high that the base of it stands above the ground. In other words, set each plant so that the base is well under the soil, but high enough that the bud is exposed. The planting hole should be made approximately as deep as the roots are long, and the soil should be lightly firmed around them and not packed so hard that the roots are injured in the process. The soil will become packed well enough after it has been watered. Berries begin ripening soon after the first of the year.

Summer Gardening

There are always a few hardy souls who insist on trying to garden all year, probably to prove that they can do it. Winter-time gardening is difficult enough, with all the pests that compete with the gardener for his harvest. In the summer pests are ten times as bad, and constant vigil is necessary if the gardener is to keep ahead of them. Moreover, only a few crops are suitable for growing during the summer in south Florida.

Probably the easiest of all crops to grow, and the one which pays the greatest dividends in summer, is okra. The principal pest is the aphid, which can be kept under control with any of several insecticides. When using insecticides on vegetable crops, directions must be carefully followed and all of the manufacturers' cautions should be observed. Okra seeds are planted directly in the rows, which are made about thirty inches apart. Thin plants to about a foot apart. The best okra is harvested when the pods are about half mature, which is one reason why home grown okra is superior to that purchased at a store. In order to be able to harvest enough okra at a time, at least thirty to forty feet of row is necessary. Harvesting can be done every second or third day over a period of several weeks. As may be surmised, one can get awfully tired of okra unless neighbors also like it.

Cucumbers, cantaloupes, and squash thrive during the summer, but to grow them well the gardener needs to spray weekly with a combination of fungicide and insecticide, especially recommended for melons. Zineb, a fungicide, and Sevin, an insecticide, have been widely used on melons. These crops are subject to stem borers, which bore into the vines, and to pickle worms, which attack the fruits before they reach maturity and spoil them. Moreover, cucumber beetles, squash bugs, and aphids also feast on melons. And in addition to the insects, mildew attacks the plants during humid weather. Only weekly spraying can avert disaster.

Gardening in the summer is usually a one-shot experience. Not even in winter do many persons continue to make a garden year after year. The availability of fresh vegetables in stores at all times, plus the vast improvements made in canning and freezing, cause most persons to wonder whether their time should not be better spent in some other pursuit. Certainly no one would make a vegetable garden on the assumption that one is saving money. Storebought vegetables are much cheaper, and, except in a few cases, superior to anything the gardener can produce in his backyard patch.

Plants like this Jasminum should be set beneath windows

Pittosporum takes pruning well at hands of master gardener

Chapter 16

PRUNING

Pruning is a year-round job in south Florida where the growth of plants never ceases. One cannot pick out a month and say that everything should be cut back at that time. Plants which bloom early in the year are best pruned in spring or early summer. Trees such as the live oak which lose their leaves during the winter may be best pruned in late fall or early winter. Large evergreen tropical trees, such as the ficus, should be pruned in late spring or late summer, in order to give them time to recover during the growing season. Then there are plants such as the poinsettia which require pruning several times between spring and the first week of September. Dead wood or diseased and weak branches may be removed at any time of the year.

Pruning should be done for a purpose. It is not something that should be done for the sake of exercise. Unless a plant requires pruning it should not be pruned. Pruning is done to give shape to a tree or a shrub and to remove weak or awkward branches. Pruning may be done to keep a shrub or tree under control, such as cutting back a hibiscus which has been unwisely planted beneath a window, or a ficus tree to keep it from swallowing up the garden and consuming the house.

Of all the activities a gardener has to learn, none requires more skillful study and application than pruning. It is at the same time a science and an art. It is helpful to know something of the physiology of plants; to understand a little about why new growth is likely to appear near the end of branches; to understand why it is necessary to cover pruning wounds with waterproof paint to prevent decay, and why cutting back a flowering shrub severely at the wrong time may induce lush growth but few if any flowers.

The best time to do drastic pruning of most plants in south Florida is near the beginning of the rainy season, in May or June. Screens and hedges which need reducing can be cut back at this time, and it is a good time to cut back big shade trees. Plants have the summer months ahead of them in which to recover. Obviously this is not the best time to cut back every kind of tree or shrub. A summer flowering tree or a mango in full fruit should receive a short reprieve from the pruning blades, until after the flowering and fruiting period is over.

Developing Character

The pruning of a plant should begin in the nursery while the plant is small. For here is where the shaping of a tree or shrub is done. A shrub should be trained as a shrub. A tree should be trained as a tree. Each plant is selected and set out in a landscape plan because it has certain desirable characteristics. Except in a few instances, shrubs are not useful as singular individuals, but are chosen because they also sit well in association with other plants. Thus when a shrub is pruned, not only must we consider its individual characteristics, but its relations with plants nearby. What is a plant to be – a tree or a shrub? And what kind of tree form or shrub form is best suited for the natural character of the plant? We should not attempt to make a formal plant out of one which naturally grows in a straggly way. It is difficult to make a low, compact plant out of a *Melaleuca leucadendron*. It would be next to impossible to make an upright plant out of a normally sprawling plant such as *Bauhinia galpinii* or one of the viney jasmines.

In every garden where good design is considered, there are many types of plants – trees, shrubs, ground covers, and grass. And these plants vary among themselves in several ways. They vary in texture, form, and color of foliage. These characteristics should be considered before the pruning tools are put to work. It is easy to destroy the natural texture and form of a shrub by running a pair of pruning shears over it. Only box type hedges, or individual semi-formal plants should be sheared. But to shear everything, every croton beneath a window, every individual hibiscus shrub planted singly for its flowers, every pittosporum or Turk's cap – well, that's carrying shearing too far. And yet this is the most common type of pruning in the Miami area.

Good Tools Important

Before a gardener can hope to do a good job of pruning a set of high quality pruning tools is required. One needs a good pair of strong side clippers, which can be used to remove small twigs or branches up to one quarter inch in diameter. These should be used frequently. Dead or weak twigs should be removed when discovered. Awkward growth should be headed back. Suckers or water sprouts, which often spring from the base of the trunk or lower branches of a citrus tree, should be removed as soon as they start.

For larger branches the gardener needs a pair of loppers, or a pruning saw, depending on the size of the branches to be removed. And where trees are to be pruned, a pruner on a long handle, with the lopper operated with the aid of a line, may be needed. Some are made with a saw on the end. Fairly large branches can be removed with these pruners twelve feet or more above one's head. One other item is needed, a waterproof paint to cover pruning wounds. Never use an oil paint. Pruning paint can now be bought in containers with which it can be sprayed onto pruning wounds.

Although constant shearing of shrubs should be avoided, it is difficult, if not impossible, to tell the gardener who mows your lawn and maintains your shrubbery how to prune. He is in a hurry to move to your neighbor's garden. If you cannot take time to do the pruning yourself, you may have to be satisfied with something less than perfect.

A good way to learn pruning is to look over the shoulder of an expert nurseryman and see how he does it. Let's take, for example, a young pittosporum. If not pruned in the nursery, this plant may send up a few tall branches, or perhaps a single branch, and may not grow into a compact shrub.

The nurseryman makes a shrub by inducing it to develop many branches – six, ten, twelve or more main stems from within inches of the ground level. This is done by heading each new branch back at the right time. For instance, the nurseryman comes to a shrub with three branches. He cuts each branch back to six inches of its origin. Fertilizer is then applied. After a few weeks of good growing conditions the nurseryman goes through his plants again. Now he finds that the plant which originally had three branches has nine. He cuts each of the nine branches back a little, applies more fertilizer and keeps his plants well watered. A few weeks later the plant which started with three branches has over twenty branches.

Plants vary in rate of growth, of course. It may take two or three years to produce an excellent shrub. That's why top grade shrubs cost more than fully grown shrubs. They require more time to grow, more labor time and more space in the nursery. The pittosporum is a type of plant which lends itself to training through expert pruning. This is an important shrub in many Japanese gardens. It can be made to take the form of a rugged old tree in miniature, or grown to sprawl. It can be trained low to serve as a ground cover. It is amazing what can be done with many plants if expertly pruned.

Inexpert pruners are reluctant to go far into a mature shrub when pruning it. They are more inclined to stay close to the tips of the branches, cutting off this one and then that one, never very far back. Then, every two or three years, the entire plant is cut back, nearly to the ground. Thus the gardener is timid on one hand, and very aggressive on the other. With better knowledge of plants and pruning, this periodic mutilation could be avoided.

Knowledge of Plants Helps

By keeping shrubs thinned out a little, one can encourage new growth from inside the plant to be developing at all times. Instead of shearing the tips of branches, individual branches are cut far back into the shrub itself. Just how far a single branch should be cut back should be carefully determined by the gardener. If it is an unhealthy branch, then it might be removed completely. But if it is removed for the purpose of encouraging the beginning of a new flush of growth from the interior of the shrub, then it must not be cut back all the way to the trunk. A stub of a branch must be left if new growth is desirable. The stub should be five or six inches long, long enough to support new shoots.

Very often it is possible to remove up to three or four branches at a time, especially in the spring or early summer, when rapid replacement of growth is assured. Such a plant, with several branches removed, may appear to be a little too open, but the air spaces are quickly filled up by space-seeking twigs from the other branches and from emerging new growth.

There is no doubt that this kind of pruning does require some careful thought. How well one does it will depend on how well one wants to do it. Each plant should be studied from time to time, while one is walking through a garden or working in it. An expert gardener will know in advance which branches should be removed. One might argue that such an approach to gardening would require a great deal of time. This is not necessarily true. The success of such pruning depends on first knowing how to approach it; secondly, on the willingness to perform and, thirdly, on having the right tools. Anyone who wants an excellent garden will be willing to spend the necessary time and energy to keep it in good condition.

The gardener approaches the pruning of a tree differently from the pruning of a shrub. Usually a single trunk is desirable, at least for six feet or more from the ground. Selection of the height to encourage branching will depend on where the tree is growing. If in a parkway between the sidewalk and street, the lower, scaffolding branches should start above head height. So should a shade tree in one's garden under which people will be walking. In groves, the branching of a mango tree may be started three or four feet above the ground. Several scaffolding branches should be encouraged.

To encourage the development of scaffolding branches, top the young tree at the desired height, or pinch out the leading bud on the single trunk. Several branches usually will form just below this point. Leave all branches which form up to five or six. If later they become too crowded, as the tree begins to take shape, one or more may be removed. If one branch tends to be stronger than the others, keep it headed back until the weaker branches catch up. Unless headed back, these vigorous branches very often will take over and cause the tree to grow lop-sided. If trees or shrubs are permitted to get out of hand, severe pruning may be needed to give them a desirable form. Growth should be equally distributed throughout the top of a tree, and pruning is required to encourage this. Problems such as these make pruning a year-round activity.

When to Apply Fertilizer

Practices vary in the fertilizing of newly pruned shrubs and trees. If only a few twigs are removed, fertilizing may be done on a regular schedule. If a plant has been trimmed back severely, an application of fertilizer as well as periodic irrigation may be needed if rapid recovery is to follow. Some gardeners fertilize a little at the time of pruning. A heavier application of fertilizer may be made after the plant has covered itself pretty well with new growth. Another practice is to fertilize a couple of weeks before pruning. Then, after the sap begins to flow, the pruning tools are put to work. The theory is that with a heavy flow of sap the plant will burst forth with new growth evenly distributed on its branches. Thus a very bushy plant is produced.

Only sharp tools should be used. If large branches are to be removed, they should be removed in sections, or a little at a time, until only a stub remains. Then remove the stub, cutting it as closely to the trunk of the tree as possible. This method usually prevents a heavy branch from splitting and leaving a ragged scar on the trunk. Short stubs or branches should always be removed, of course, as they offer ideal places for decay organisms to enter and to set up rot.

Choose pest-free plants like this agave whenever possible

Pest-free Schefflera makes an excellent screening plant

Chapter 17

INSECT PESTS AND DISEASES

The south Florida gardener must engage in a year-round battle with insects and plant diseases. Insects thrive and multiply during the mild winters. Moisture conditions are favorable for the growth of disease-causing fungi at nearly all times. Thus the gardener who wishes to grow quality plants must be prepared to spray or to dust with insecticides or fungicides every month in the year. But in order to maintain control over insects and diseases, some knowledge of them and how to recognize them is necessary.

Florida has ten thousand kinds of insects. Fortunately, relatively few of them do major harm. Many insects are beneficial because they prey on insect pests. The lady beetle is an example of an insect which preys on other insects. There are several kinds. Probably the best known lady beetle in Florida is a turtle-shaped bug about one-fourth of an inch long, with two crimson spots on its ebony back. It feeds on aphids and on certain scales. Unfortunately, the lady beetle larva looks like a small hideous caterpillar rather than like a lady beetle. Gardeners often mistake it for a pest and spray it with insecticides. A closer look would reveal that the lady beetle larva does not feed on plant tissue but on other insects. Other beneficial insects are certain types of wasps, bees, and harmless flies which one often sees about the flowers of fruit trees early of a morning. These insects are important in the pollination of flowers, thus aiding in production of fruit.

Spraying should always be limited to the control of harmful insects which are actively engaged in the destruction of plants.

It is hardly possible for anyone but an entomologist to be able to recognize all the harmful insects. But if one learns the insects by groups, such as scales, caterpillars, aphids, and bugs, it should be possible to keep pests under control without a great deal of trouble, by spraying with the insecticide recommended for the group of insects found to be doing damage. Learning to identify insects this way should not be difficult. A great deal of literature is available on them, including bulletins published by the Agricultural Extension Service of the University of Florida and by the U. S. Department of Agriculture in Washington. In recent years many inexpensive paper-back books have come out which picture insects in color and describe and give their life cycles. If the gardener is unable to get an insect identified locally, he can send it to the Florida State Plant Board, Gainesville, Fla.

Scale Insects

One of the most difficult groups of insects the gardener has to deal with are the scales. They usually go unnoticed until after they have reached the age when they are permanently attached to leaves. Most of them have scale-like coverings, and when one is gently pressed by means of a thumb nail, fluid exudes from beneath the shell along its edge as the insect is crushed. Scale insects are active when young, crawling about on the leaves and stems of plants, but the gardener seldom sees them while they are in this stage, unless he becomes especially interested in insects and is constantly examining his plants with the aid of a hand lens.

Not all scales are immoble when mature. Cottony cushion scale and mealybug move on plants after reaching maturity and do not attach themselves permanently to a site as do most other scales. These two scales have a great many enemies, however, including several small wasps and flies, in addition to the lady beetle. They are usually fairly well kept under control in older gardens where there has developed a kind of balance of power between the insects which live on plants and the insects which live on other insects. However, cottony cushion scale and mealybug can become severe on plants grown indoors or within screened patios. Thus constant attention should be given to all ornamental plants grown under conditions where pest insects enjoy protection from their enemies.

Only a few insecticides will control scale insects. Oil emulsion is an old reliable method, and it is still one of the best agents for the control of some armored scales, and especially Florida red scale, a common pest of fruit trees and ornamentals. Malathion does a good job of controlling most scale insects, but usually does a better job when it is combined with oil emulsion. This combination is available at garden supply stores.

Commercial growers usually use parathion for control of scale insects in groves, but this material is considered too dangerous for use by the homeowner, unless he has special protective gear for applying this very toxic phosphatic compound. Where mealybug or cottony cushion scales are found on indoor plants, these ordinarily can be controlled by cleansing the plants with a moist soft cloth.

Chewing Pests

The gardener will have much less trouble identifying such pests as caterpillars, grasshoppers, stink bugs, slugs, and beetles. Some caterpillars, however, are night feeders, such as the sod webworm which causes widespread damage to lawns during the warm season. When night-time feeders are at work, it is usually a detective problem to identify the type of insect by the signs left. Caterpillars chew on foliage and they leave their droppings. Slugs leave slimy trails.

The grasshopper usually comes in spring and early summer. Young grasshoppers sometimes appear in large numbers, threatening to consume every living thing. The young, or nymphs as entomologists know them, are without wings, and are sluggish and awkward. Even the most casual observer should be able to recognize them as grasshoppers. They are voracious feeders, eating continually until they reach maturity. They like especially to feed on the leaves of amaryllis, crinums, and daylilies, but they also will feed on any tender foliage. One does not obtain a quick kill with the best of insecticides. A more practical way to kill grasshoppers is to knock them off plants and to crush them on the ground. A grasshopper-killing insecticide should be applied, however, whenever these pests appear in very large numbers. The material should not only be applied to plants on which grasshoppers are feeding, but also on the grass within the vicinity. Chlordane and toxaphene usually give good control of grasshoppers.

Caterpillars are easier to kill. A number of insecticides give good control. Spraying should be done as soon as the first sign of caterpillar damage is noticed. This is especially important for the control of armyworms and sod webworms. For instance, DDT and lindane give excellent control of sod webworms when these caterpillars are quite small, but when they become larger they become more difficult to kill, and as they reach full size the gardener may have to turn to toxaphene to get good control. Spraying for control of the giant hornworm, which is a constant hazard to the growing of poinsettias, is likely to be impractical. Only one or two of these insects are found at a time. The best control is to knock the hornworm off the plant on which it is feeding and crush it, as is recommended for the control of grasshoppers.

Identifying Slugs

Caterpillars are often blamed for the disappearance of leaves from our best ornamentals, when the culprit is not an insect at all but a kind of snail. The slug, a snail without a shell, and reaching two inches in length, is a major garden pest in south Florida. It is a night-time feeder and is active while the gardener sleeps. On the following day when the gardener notices that the leaves are disappearing from ground covers and other plants with tender foliage, he is likely to apply an insecticide, believing the damage to have been caused by caterpillars. But the slimy slugs are not affected by insecticides. The only way to control slugs is to use a bait containing metaldehyde. Slug bait, sold under trade names, may be obtained from garden supply stores. Apply to the soil about affected plants late in the day. Moisten very slightly by passing a light spray from a garden hose over the area. On the following morning any remaining bait may be washed into the soil, to keep birds from eating it. Small animals, especially dogs, will eat slug bait if it is applied where they can get to it. Slugs live in moist, shady places, and this is where the bait should be applied, where plants are dense. Slugs hide beneath rocks, in leaf mulches or other dark places during the day time. Their slimy trails are a give-away. Fortunately, slug bait is very effective, and one application is usually sufficient to wipe out most of the slug population. Two or three additional applications, made at two-week intervals, should get the slugs which hatch after the first application.

Aphids and Thrips

Aphids are easily detected without the aid of a hand lens. However, they often go unnoticed until they have done severe damage. Aphids, or plant lice, as they also are known, are usually found on new growth, and especially on the new growth of citrus, gardenias, or ixoras. Inexperienced gardeners may notice that the leaves of their citrus trees have become distorted or curled by some mysterious agent, and once again they may mix up insecticides and spray. But by this time the damage has been done, and if the leaves are mature at this stage, the aphids also may have vanished. The best way to control aphids is to examine new leaves as they come out, and apply an insecticide such as nicotine sulphate, malathion, or lindane if aphids are found.

Thrips are very small soft-bodied, yellowish insects which are likely to escape notice of the gardener who does not have a hand lens to examine suspicious signs on his plants. These pests attack leaves and flowers. The red-banded thrips is a serious pest of the mango. Fortunately, thrips can be controlled with most of the newer insecticides, including malathion, lindane, chlordane, or sevin.

Spider Mites

Plant mites are smaller than thrips. One of the largest and most common of the mites is the red spider, about one-sixtieth of an inch long. It can be easily seen with the aid of a ten-power lens; but young, sharp eyes may be able to detect it without the aid of magnification. It is truly red, and when examined closely, one can see why it has the name of red spider, for it resembles a spider, of which it is a relative. There are many kinds of spider mites which attack plants, but they are not necessarily red in color. Some are green. Some may be nearly transparent. Others may be yellowish-brown. Not all can be seen with the aid of a hand lens. Some are so small that a low-power microscope is required to see them.

The presence of mites is usually known by the damage they do. They puncture leaves and suck juice from them. When they are permitted to accumulate in large numbers they can do a great deal of damage, causing plants to lose their foliage. Mites may be especially severe on indoor plants, and even to orchids. But mites can be active anywhere when conditions are right. Various kinds of mite feed on citrus, avocados, roses, palms, and many of the common foliage plants. Close checking of plants infested with mites will show numerous stipples and yellow spots where they have concentrated. Most of the ordinary insecticides do not control these pests. A number of miticides are on the market which give good control without killing beneficial insects that help to keep the mite population down. Mites are usually more severe in gardens where several kinds of insecticides are used indiscriminately. A number of insects are predatory on mites and for this reason the gardener should never spread insecticides indiscriminately throughout the garden. Such spraying destroys the delicate balance of nature and permits a rapid build-up of plant consuming pests, especially mites, after their enemies have been killed. For it must be realized that to kill every insect is next to impossible, and there will always be a few left to rebuild the population. Pest insects seem to develop much more rapidly than do predatory insects. At least in the beginning they do. But by wise use of insecticides, the gardener can enlist the help of the many predators in the constant battle against his insect enemies.

Rare Insects

Various kinds of chewing bugs, juice-sucking bugs, and caterpillars sometimes appear in large numbers to attack plants. The gardener may have never seen the insects before. He may wonder if it is a new kind of pest from a foreign shore brought in accidentally. The population builds up rapidly, and the gardener begins to believe that the insects are about to take over. He hastens out to get the insect identified and to buy insecticides, but perhaps before he has time to spray, or even to get the insect identified, the pest has disappeared. If the insect specimen is sent to the Florida State Plant Board the gardener may learn later that it is a bug with a name a mile long, and described as being of little economic importance, meaning that it has a low rating as a plant destroyer. The ways of insects, and their comings and goings, are not understood by the experts. So the gardener need not be baffled over what he might see from day to day about his plants. If pests build up in large numbers, he usually can get adequate control of them by spraying with malathion, lindane, or sevin.

Specialized insect problems are discussed under chapters on ornamentals, vegetables growing, citrus, tropical fruits, and lawn grasses.

Plant Diseases

Diseases in plants are caused by fungi, bacteria, viruses, and also by mineral deficiencies. Although mineral deficiency does not come under the heading of pests, the gardener nevertheless must be able to recognize the symptoms as being distinct from fungus or bacterial diseases. There are, however, few gardeners who could tell the difference between the symptoms of virus disease and the symptoms of iron or zinc deficiencies which show up in the leaves of plants.

Whether the gardener ever learns how to recognize these differences will depend on how far he wishes to go in becoming a master horticulturist. The gardener learns to recognize plant trouble from experience. Spots on leaves, wilting, damping off of young seedlings, and the decaying of twigs tell us that something is obviously wrong. Leaf troubles can be caused from sun-burn, from winter injury, from spray or fertilizer burn, or by mites feeding on the under side of leaves. Sometimes strong chemicals, either fertilizer or spray materials, accumulate at leaf tips and cause them to turn brown. Wind also may cause tip-burn, especially on the leaves of young plants which are not well established. The root system of a poorly established plant may not be extensive enough to take up moisture as rapidly as it is given off by the foliage. This may cause leaf cells to become dry and to die.

Applying insecticides or fungicides when injury is caused by mechanical means or as a result of strong chemicals does more harm than good. For this reason the gardener should know what is causing the trouble. Fungus or bacterial spots are usually quite uniform in appearance. Often the center of a fungus-caused spot will be lighter in color than the margins. Bacterial spots may appear to be water-soaked and are sometimes translucent, and may be covered with a slime. It is hardly possible to prevent fungus and bacteria from attacking all plants. In time the gardener learns that about the best he can do is to take care of the more serious or specialized problems among his plants, such as anthracnose on mangos, blight on tomatoes, black spot on roses, mildew on annuals, sooty mold on citrus, ixoras, and gardenias.

Anthracnose shows up as black spots on leaves and fruits of mango trees, but its damage to flower panicles may take place before any injury is noted. Thus spraying with a fungicide as the flower panicles begin to open, as recommended in the chapter under *Tropical Fruits,* is a necessary precaution. Preventive spraying of tomatoes against late blight is a must for a back-yard gardener as well as the tomato farmer. Sooty mold fungus, however, is not a plant disease, although it appears to be. This fungus lives on honeydew excreted by scales and certain other insects. After the insects are controlled the sooty mold fungus eventually disappears.

Ordinarily the gardener does not have to worry about bacterial diseases. But should they attack his plants, as, for instance, his fancy greenhouse plants, he can use an antibiotic spray. The ordinary fungicide are not efficient in controlling most bacterial diseases. Antibiotics, on the other hand, also give fair control of many fungus diseases, including some of the difficult ones which attack lawn grasses.

Virus Difficult to Control

No amount of spraying with antibiotics or fungicides will have any effect on virus diseases or correct nutritional troubles. When plants are attacked by virus the leaves become yellow or mottled with green and yellow patterns which explains the use of the word mosaic as a synonym for virus disease. Virus-infected plants lose vigor, slow down in growth, and the leaves, flowers, and fruits tend to be poor. A number of vegetables and annuals become infected by viruses. Insects, chiefly aphids, transmit them. But they may also be transmitted through budding, as in citrus. To prevent transmission of virus through budwood, the Florida State Plant Board has set up a strict inspection service for certified nurseries which come under state regulations. A virus disease in the papaya, believed to be transmitted by aphids, makes it difficult to grow this plant successfully in south Florida. The only method of control, is to pull out affected plants before the disease is transmitted to healthy ones.

Yellowing of leaves may be caused by deficiencies of iron, zinc, or manganese. Foliage also may turn yellow and plants become sickly if one attempts to grow them in poor soil without the aid of fertilizer. Nematodes feeding on plant roots and preventing the plants from taking up sufficient nutrients from the soil also may cause foliage yellowing. A plant which requires an acid soil, such as the gardenia, may turn yellow if planted in limestone or marl soil. On the contrary, many plants which thrive best in a soil which is nearly neutral will turn yellow if planted in highly acid soil, such as is found in the low flatwoods.

To explain how the multitude of plant troubles can be identified is impossible. A thick volume is required to explain and illustrate the insects and diseases which attack citrus alone. Turf grass experts have difficulty in identifying the diseases which attack grasses. One must have several years of experience in order to be able to recognize the most common insect pests and diseases, and to know what to do in order to control them. When buying insecticides or fungicides it is helpful to take affected leaves to the garden store where one trades and try to get the trouble identified. If it is not possible locally to get positive identification, the affected leaves may be sent to the State Plant Board. By using such a system the interested, wide-awake gardener becomes acquainted with the common pests and is able to take immediate action when they attempt to take over his choice plants.

One of the surest ways to obtain good control of plant troubles is to learn all we can about the various ornamentals and fruit trees which we plant in our garden. Every gardener should know that a gardenia cannot be planted on its own roots in limestone soil, but must be budded onto the root stock of a plant which is tolerant of alkaline soil; that citrus requires expert care, including a nutritional spray at least once a year; that roses are subject to blackspot, and must be sprayed or dusted with a fungicide as a preventive measure; that the papaya is subject to a virus disease, and that affected plants must be taken out immediately; that grass is likely to be attacked by chinch bugs, caterpillars and diseases; that tomatoes, peppers, and many ornamental plants are subject to severe damage from nematodes, and that these eel-like worms may be controlled by applying a nematicide to the soil.

If the gardener realizes that the best entomologists, plant pathologists, and nematologists cannot in a life time learn to identify all the troubles to which a plant can succumb, then perhaps he may realize how difficult it is for the untrained person to know everything about the plants in his garden, and about the enemies which attack them.

Spraying Equipment

How a chemical is applied can be just as important as determining the right kind of chemical to use for a given situation. When an insecticide is used on a lawn for control of chinch bugs a large amount of water is also required. Thus the insecticide may be efficiently applied with the aid of a garden hose spray attachment. The idea is to get the required amount of the insecticide over a given area of grass, and to wash it into the grass where the chinch bugs are feeding. But to use the same type of spraying equipment for applying weed killers, fungicides, or an oil emulsion for the control of scale insects on a fruit tree may not be advisable.

Careful dilution is necessary for the selective weed killers for lawn grasses, and these must be used with extreme care if injury to the turf grass itself is to be avoided. The material should be applied on the leaves of weeds just enough to cover them but without any excess run-off. Flooding the lawn with weed killer is not necessary, and can be dangerous. Severe damage to grass may result. When fungicides or insecticides are used, careful coverage of the effected areas, or the insects themselves, is a prime necessity. It is also necessary that the materials be used at exactly the strength recommended on the label of the container.

Efficient spraying can best be accomplished by using a compressed air sprayer, the smallest of which comes in one and one-half gallon size. Usually the smallest sprayer is sufficient for a weed killer, and, once used for this purpose, should be reversed only for this weed killer. Other pressure sprayers range from three gallon sizes to twenty-five gallons, and the larger ones have motor power.

For the average size garden four sprayers are usually sufficient to cover every purpose. These would include a simple hose attachment sprayer for insecticides for lawn pests control; a one and one-half gallon pressure type sprayer for selective weed killer; a three-gallon pressure type sprayer for various insecticides for shrubbery and small fruit trees, and a trombone type sprayer for insecticides and fungicides on large trees such as mangos. The trombone sprayer is fed out of a bucket or tub. All spraying equipment should be thoroughly washed after each use.

Insecticides come in various forms, as dust, wettable powder, emulsions, aerosols, and granular. The gardener is likely to have less trouble with his spray equipment if he sticks with the emulsifiable concentrates, but these must be mixed with considerable care to prevent injury to plants. Strong solutions can cause severe burning. Application of any chemical to plant foliage at a time when the soil is extremely dry and the plant is in a state of wilt may also cause severe injury, even though the chemical may be mixed according to the manufacturer's direction. It is best to irrigate the soil well the day before chemicals are sprayed onto ornamentals and fruit trees. Granular insecticides are popular for the control of ants, mold crickets, cut worms, and other types of insects that live on the soil surface or just beneath the surface. They are sometimes used for the control of lawn pests, including the chinch bug.

The gardener should always read carefully the label on the container of any chemical he is about to apply. All insecticides are toxic. Some of them can become extremely dangerous when improperly handled. Insecticides, fungicides, and other chemicals the gardener uses should be stored in a locked cabinet which children cannot open.

BIBLIOGRAPHY

BARRETT, MARY F., *Common Exotic Trees of South Florida,* University of Florida Press, Gainesville, 1956.

BRUGGEMAN, L., *Tropical Plants and their Cultivation,* Thames and Hudson, London, 1957.

BOR, N. L., and RAIZADA, M. B., *Some Beautiful Indian Climbers and Shrubs,* Bombay Natural History Society, Bombay, India, 1954.

CHANDLER, W. H., *Evergreen Orchards,* Lea & Febiger, Philadelphia, 1958.

DORN, MABEL, *Tropical Gardening for South Florida,* South Florida Publishing Co., South Miami, 1952.

ECKBO, GARRETT, *The Art of Home Landscaping,* F. W. Dodge Corp., New York, 1956.

ENGLE, DAVID H., *Japanese Gardens for Today,* Charles E. Tuttle Co., Rutland, Vt., 1959.

FAIRCHILD, DAVID, *The World Grows Round My Door,* Scribner's Sons, New York, 1947.

GRAF, ALFRED B., *Exotica II,* Roehrs Co., Rutherford, N. J., 1959.

KENNARD, WILLIAM C., and WINTERS, HAROLD F., *Some Fruits and Nuts for the Tropics,* Miscellaneous Publication No. 801, U. S. Government Printing Office, Washington, 1960.

KUCK, LORAINE E., and TONGG, RICHARD C., *Hawaiian Flowers and Flowering Trees,* Charles E. Tuttle Co., Rutland, Vt., 1958.

KUCK, LORAINE, and TONGG, RICHARD, *The Modern Tropical Garden,* Tongg Publishing Co., Ltd., Honolulu, 1955.

LEDIN, R. BRUCE, *Palm Handbook,* American Horticultural Society, Washington, 1960. (Distributed locally by Fairchild Tropical Garden, Miami.)

MENNINGER, EDWIN A., *What Flowering Tree is That?* The Stuart Daily News, Inc., Stuart, Fla., 1956.

MORTON, JULIA F., and LEDIN, R. BRUCE, *400 Plants of South Florida,* Text House, Coral Gables, 1952.

MORTON, KENDAL and JULIA, *Fifty Tropical Fruits of Nassau,* Text House, Coral Gables, 1946.

MACMILLAN, H. F., *Tropical Planting and Gardening,* The Macmillan Co., New York, 1935.

NEAL, MARIE C., *In Gardens of Hawaii,* Bernice P. Bishop Museum, Honolulu, 1948.

ORTLOFF, H. S., and RAYMORE, H. B., *The Book of Landscape Design,* M. Barrows & Company, New York, 1959.

OAKMAN, H., *Gardening in Queensland,* Jacaranda Press, Brisbane, 1958.

STURROCK, DAVID, *Fruits for Southern Florida,* Southeastern Printing Co., Stuart, Fla., 1959.

SMILEY, NIXON, *Florida Gardening Month by Month,* University of Miami Press, 1957.

WATKINS, JOHN, and WOLFE, HERBERT S., *Your Florida Garden,* University of Florida Press, Gainesville, Fla., 1954.

WEST, ERDMAN, and ARNOLD, LILLIAN, *Native Trees of Florida,* University of Florida Press, Gainesville, Fla., 1956.

Index